Biopolymers Towards Green and Sustainable Development

Authored by

Sudarshan Singh
Department of Pharmaceutical Science
Faculty of Pharmacy, Chiang Mai University
Chiang Mai
Thailand

&

Warangkana Chunglok
School of Allied Health Sciences
Walailak University
Nakhon Si Thammarat
Thailand

Biopolymers Towards Green and Sustainable Development

Authors: Sudarshan Singh and Warangkana Chunglok

ISBN (Online): 978-981-5079-30-2

ISBN (Print): 978-981-5079-31-9

ISBN (Paperback): 978-981-5079-32-6

First published in 2022.

need for a court order if at any point you breach any terms of this License Agreement. In no event will any delay or failure by Bentham Science Publishers in enforcing your compliance with this License Agreement constitute a waiver of any of its rights.

3. You acknowledge that you have read this License Agreement, and agree to be bound by its terms and conditions. To the extent that any other terms and conditions presented on any website of Bentham Science Publishers conflict with, or are inconsistent with, the terms and conditions set out in this License Agreement, you acknowledge that the terms and conditions set out in this License Agreement shall prevail.

Bentham Science Publishers Pte. Ltd.
80 Robinson Road #02-00
Singapore 068898
Singapore
Email: subscriptions@benthamscience.net

BENTHAM SCIENCE

CONTENTS

FOREWORD

There has been growing concern about the negative impacts of environmental pollution from fossil fuels, waste from petroleum products, and non-biodegradable materials. A lot of research has been done into exploring another alternative to petroleum-based products which would be renewable as well as biodegradable and thus pose a lesser risk to the environment. Biopolymers are one such possible solution to the problem because they are typically biodegradable materials obtained from renewable sources. Moreover, biopolymers are produced by the cells of living organisms consisting of monomeric units that are covalently bonded to form large molecules. Some of the first modern biomaterials made from natural biopolymers include rubber, linoleum, celluloid, and cellophane. However, due to growing ecological concerns, the application of biopolymers enjoys renewed interest from the scientific community, the industrial sectors, and even other allied sectors.

Biopolymers towards green and sustainable development provide an up-to-date summary of polymeric materials characterized by biodegradability and sustainability. The book includes a thorough breakdown of the vast range of application areas, including pharmaceutical and medical, packaging, textile, biodegradable plastics, green synthesis of metallic nanoparticles, and many more, giving engineers critical materials information in an area that has traditionally been more limited than conventional inactive ingredients. This book aims to fulfill the current need of the researcher by providing an excellent bibliometric meta-analysis on bio-based polymers indicating potential collaboration between country, organization, institution, and authors. Moreover, a meta-analysis provided a view of recent ongoing trends in biopolymers.

I have, no doubt, that this book will be well-received by all those in the pharmaceutical and agro-industry, academia, and other research organizations who continually seek inactive functional biomaterials for improved drug formulation and development, especially scientists and students working with biopolymers.

Chuda Chittasupho
Faculty of Pharmacy
Chiang Mai University
Thailand

PREFACE

Biopolymers are polymers synthesized by living organisms. They can be polynucleotides, peptides, or polysaccharides. These consist of long chains made of repeated, covalently bonded units, such as nucleotides, amino acids, or monosaccharides. Cellulose is the most common organic compound, and about 33% of the plant matter is cellulose. Biopolymers can be sustainable and carbon neutral and are always renewable because they are made from plant materials that grow indefinitely. These plant materials come from agricultural non-food crops. Therefore, the use of biodegradable polymers creates a sustainable industry. In contrast, the feedstock for synthetic polymers derived from petrochemicals will eventually deplete and most of them are non-biodegradable. Non-biodegradability issues of synthetic inactive pharmaceutical ingredients strongly emphasized innovators towards the development of biopolymers. Recently natural biodegradable excipients gained significant attention due to their sustainability and engineered applications. Innovative technologies to transform these materials into value-added chemicals *via* novel graft-polymerization or co-processing techniques for the production of high-performance multifunctional and low-cost polymers with tunable structures are key parts of its sustainable development. Besides, the development of state-of-the-art advanced characterization techniques for these engineered materials is an essential component in uncovering their specific structure and facilitates the application of these materials in the new research area. This expansion is driven by a remarkable progress in the process of refining biomass feedstock to produce bio-based building blocks. The book has been written to provide a broad platform for innovators and researchers in the area of biopolymers' development with major biomedical and agro-industrial applications. Furthermore, to communicate the state-of-the-art work related to the transformation of natural materials into value-added pharmaceutical inactive ingredients, a brief on the modification and fabrication of new biopolymers, and their characterization including the application in the textile and plastic industry has been emphasized. Moreover, the book presents updated information and addresses various issues on emerging new sources of biopolymers with multifunctional efficacy, food, and drug administrative regulatory requirements, with their impact on the ecosystem and human health. Additionally, the book also provides updated information on a meta-analysis of bio-based pharmaceutical excipients.

There are numerous books about biopolymers covering the scientific research that is enabling the new generation of degradable plastics. The goal of this handbook is to bring together some of the core knowledge in the field to provide a practical and wide-ranging guide for engineers, product designers, and scientists involved in the commercial development of biopolymers and their use in the various biomedical, environmental, and agro applications. Additionally, information on the impact of non-biodegradable materials on human health and the environment has been taken into consideration. This book gives a brief account of inactive ingredients originating from plants and their characterization techniques with pharmacokinetics. The book also covers a summary of the bibliometric meta-analysis of bio-based polymers.

We acknowledge Walailak University for extending the library facility and providing access to Scopus. Moreover, Dr. Ozioma F Nwabor, Division of Infectious Diseases, Department of Internal Medicine, Faculty of Medicine, Prince of Songkla University, Hat Yai, Songkhla, Thailand is acknowledged for the valuable suggestions and critical comments.

CONSENT FOR PUBLICATION

Not applicable.

CONFLICT OF INTEREST

The author declares no conflict of interest, financial or otherwise.

ACKNOWLEDGEMENTS

Declared none.

KEY FEATURE

• Provides an up-date summary on recently discovered natural polymeric materials

• Recently discovered new sources of biopolymers have been presented in this book.

• Presents a thorough breakdown of the vast range of application areas including fabrication of conventional and novel drug delivery, polymeric scaffolds, composites, microneedles, and green synthesis of metallic nanoparticles.

• Bibliometric meta-analysis indicating potential collaboration between country, organization, institution, and authors with a view on recent ongoing trends with biopolymers.

• A summary of pharmacology and pharmacokinetics on the inactive pharmaceutical ingredient presented

Sudarshan Singh
Department of Pharmaceutical Science
Faculty of Pharmacy, Chiang Mai University
Chiang Mai
Thailand

Warangkana Chunglok
School of Allied Health Sciences
Walailak University
Nakhon Si Thammarat
Thailand

CHAPTER 1

Overview on Bio-based Polymers

Abstract: Synthetic polymers are an imperative manmade discovery that has long been under environmental scrutiny due to their several detriments such as slow or non-degradation, diminutive re-usage, and severe milieu effects. A rough estimate indicates that 8300 million metric tons of virgin plastic are produced using synthetic materials to date, of which only 9% have been recycled until 2015. The detrimental effects of a synthetic polymeric waste product on surroundings can be slowed down by replacing it with biopolymers. Biodegradable polymers are materials that degrade due to the action of either aerobic or anaerobic microorganisms and/or enzymes. Environmental protection agency and PlasticEurope indicated that biodegradable polymers have shown a promising impact on the environment with a decline in the waste and toxic gas produced by either burying or incinerating synthetic polymers and their products. Moreover, the replacement of plastic products with bio-polymeric material for general, pharmaceutical, and agricultural use has also shown a significant decline in waste plastic in landfills and oceans. Furthermore, the potential market share of biopolymers growing gradually and is projected to generate 10.6 billion US Dollars by 2026. However, the potential biodegradable polymers market capital share has yet not reached its peak, due to the non-availability of specific regulatory standards and approval process. Thus, a complete replacement of synthetic polymers with biodegradable polymeric materials can be a paradigm shift for nature and human beings. This chapter acmes on the history of biodegradable materials and their impact on nature with their regulatory requirements to gain market capital share.

Keywords: Biodegradable agro-materials, Biodegradable material, Biopolymer market, Biopolymers.

INTRODUCTION

Mankind was familiar with bio-based materials and their use since the beginning of civilizations. The synthetic polymer industry in its initial stage assured systematic preservation of the environment with progressive support to human beings. However, with the discovery of fossil fuels for the synthesis of petroleum-based polymers and their products, the development and innovation of natural polymers suffered major setbacks. Moreover, the innovation of single-use synthetic polymer-based plastic materials severely affected the ecosystem. In view of the potential disadvantages of synthetic polymers towards the environment, biodegradable polymer regained consideration among researchers, pharmaceutical manufacturers, and other allied industrial sectors. Biopolymers are polymeric materials synthesized by living organisms that can be polynucleotides, and polypeptides, or polysaccharides. Biopolymer mainly consists of long chains made of repeating covalently bonded units, such as nucleotides, amino acids, or monosaccharides. Furthermore, biodegradable polymers have received significant attention in the last few decades due to their various potential application including the development of novel dosage forms, fabrication of agro-biotechnological products, *etc*. The biodegradability of such polymers or polymeric materials results in the formation of by-products such as CO_2, N_2, H_2O, biomass, and inorganic salts upon breaking down either by aerobic or anaerobic microorganisms. However, this degradability does not include polylactides polymers that hydrolyze comparatively at a higher rate even at room temperature and neutral pH in the absence of hydrolytic enzyme. Moreover, biodegradation does not mean that all material can be processed into compost or humus. In addition, biodegradation significantly differs from the bio-erosion process. Bio-erosion is a process of conversion of initially water-insoluble material to a slow water-soluble material that may or may not involve any major chemical degradation. Polymers are versatile compounds and are classified based on several parameters such as the source of availability, type of polymerization, monomers in the repeating units of polymers, molecular forces, *etc*., as presented in Fig. (**1**). In this chapter, an overview is presented on the history of biopolymers and the impact of synthetic polymers on the ecosystem. Furthermore, a brief account of the regulation involved in the safety and efficacy of biopolymers and their market potential concerning the maintenance of the carbon cycle within the environment has been taken into consideration, with special attention to the use of the polymer-based product in the pharmaceutical and agroindustry.

HISTORY OF BIODEGRADABLE POLYMERS

The term excipient is derived from the Latin word, *excipiens*, which means either to receive, to gather, or to take out. The definition of excipients has changed from time to time with its functions. The International Pharmaceutical Excipients Council (IPEC) defines an excipient as any substance other than the active drug or prodrug that is included in the manufacturing process or is contained in the finished pharmaceutical and relevant products [1]. Several incidences including phenytoin toxicity in 1968 and lack of strict regulations raised serious concerns and steered IPEC to mandate the manufacture for providing material safety data directly or indirectly consumed. The synthetic polymer market is growing exponentially and has become an integral part of day-to-day human life due to the enormous use of polymer-based products. These polymers are out product of petroleum oil industries or chemically synthesized *via* polymerization of several monomers. The market available petroleum-derived synthetic polymers are designed to resist the biological attack and stabilized with antioxidants and heat stabilizers that protect them from environmental degradation. Furthermore, synthetic polymers' production was significantly accelerated by a global shift from reusable to single-use plastics and surpassed most manmade products. In consequence to that the share of plastics in municipal solid waste increased from less than 1% in 1960 to more than 10% by 2005 [2]. Although products made using synthetic polymers are a more economically feasible choice than biodegradable polymers, however, the scenario has changed as such synthetic polymers produce detrimental effects to the environment and health of several organisms on enduring use.

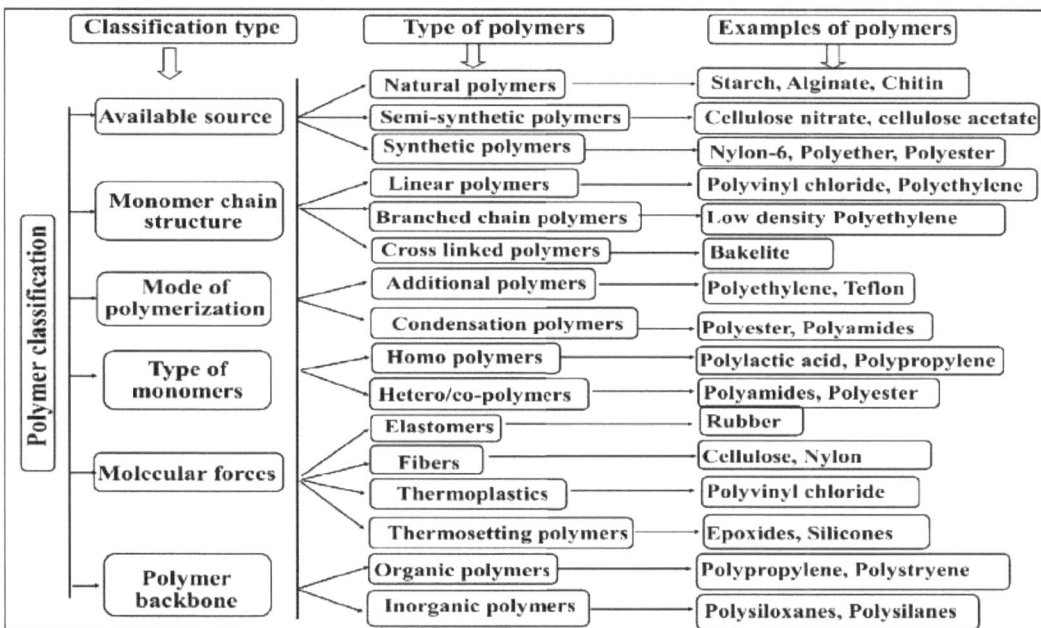

Fig. (1). Classification of polymers [4].

Decreasing the use of non-biodegradable polymers and reducing the solid waste generated from them have become a high priority due to the rising cost of petroleum oils with increasing concern about the preservation of ecological systems. In addition, the use of synthetic polymers generates substantial environmental pollution and damage to wildlife. Additionally, incineration of the synthetic polymer-made product presents serious environmental issues due to toxic emissions including dioxins, furans, mercury, and polychlorinated biphenyls [3]. Moreover, toughen legal requirements of several countries for the management of waste caused a concern to focus on the expansion of biodegradable functional polymers. For such different issues, it is necessary to replace the synthetic polymer and its products partially or completely with biodegradable materials that can degrade with time. Furthermore, while selecting and using the renewable biomass or biodegradable material, a complete understanding of the carbon cycle is required. The carbon cycle is a complex process in which carbon is exchanged between the major reservoirs of carbon within the planet. The imbalance in the carbon cycle with a rapid release of CO_2 that could not be completely compensated *via* the photosynthesis process leads to global warming (Fig. **2**). Therefore, a biodegradable material is required that not only replaces the synthetic polymers but must help in re-balancing CO_2 in the environment.

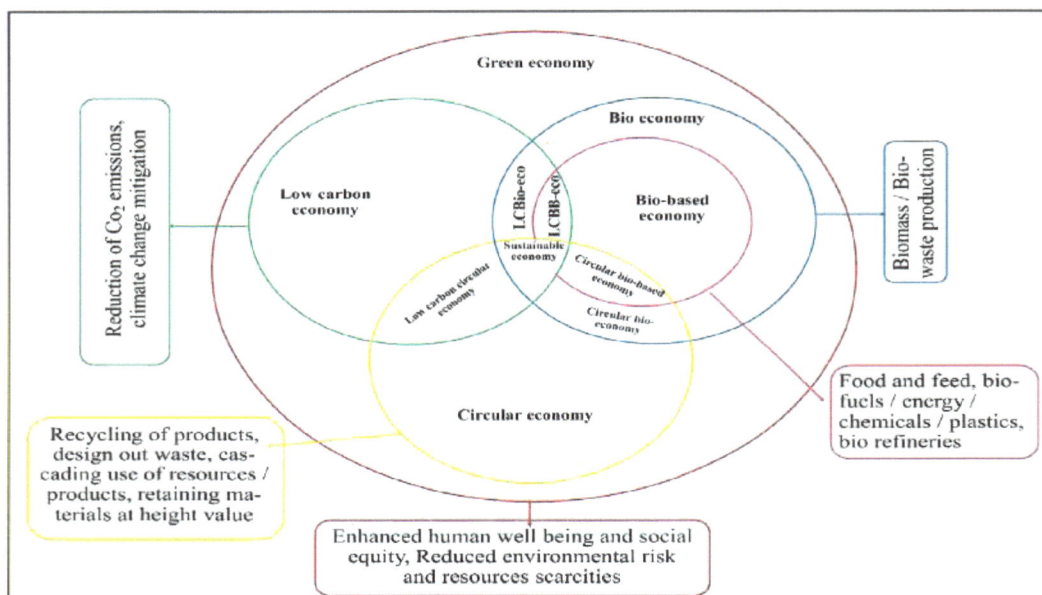

Fig. (2). Suppressed carbon emission with deceiving green economy, surrounding circular economy, bio-economy, bio-based economy, and low carbon economy, with advantage offered by each economy [4] (For interpretation of the results to color in this Fig. legend, the reader referred to either web version of this chapter or color print).

The first replacement of a synthetic polymer and its product fabricated using petrochemicals has been identified as catgut sutures made of biodegradable materials, which dates back to at least 100 AD [5]. Although the first sutures were fabricated using the intestine of sheep, the modern sutures are manufactured using purified collagen extracted from the small intestine of cattle, sheep, or goats [6]. Moreover, the first manufactured bioplastic was prepared in 1862 by Alexander Parkes using Parkesine, latter in 1897 and 1930 a biodegradable plastic was fabricated using casein from milk and soybeans, respectively. Whereas the commercialization of biodegradable plastics and polymers in the market was first introduced in 1980 [7], however the exclusive biodegradable polymers received significant attention in 2012, when Geoffrey Coates of Cornell University, New York, United State of America received the Presidential Green Chemistry Challenging Award for developing green commercial biodegradable polymers. In recent years, there has been a marked increase in interest in the use of biodegradable materials including packaging, agriculture, medicine, and other areas. Several biodegradable polymers commonly known as starch or raw carbohydrates originating from fruits seeds and fiber extracted from natural resources have gained significant attention. The belief is that biodegradable polymer materials will reduce the dependency on synthetic polymer at a low cost, thereby producing a positive effect both environmentally and economically. Moreover, the current trends in biodegradable polymers indicate noteworthy developments in terms of unique design strategies and engineering that could offer advancements in polymers with excellent performance. However, until now, natural biodegradable polymers have not found extensive commercial applications in pharmaceutical and food-packaging industries to replace the conventional synthetic adjuvants which might be due to shortfalls in either technology transfer or production cost.

Biodegradable materials are those substances whose physio-chemical characteristics completely deteriorate and degrade when exposed to microorganisms, aerobic or anaerobic process, resulting in the generation of natural byproducts. However, the biodegradability of polymers depends on several factors including the surface area, molecular weight, glass transition temperature, melting point, and crystal structure. Biodegradable polymers naturally and synthetically made, consist of ester, amide, and ether functional groups. Moreover, the biodegradable polymer is often synthesized by ring-opening polymerization, condensation reactions, and metal catalysts. Furthermore, biodegradable polymers are classified according to their origin and synthesis method, composition, processing method, economic importance, application, *etc*. Fig. (**3**) shows the various applications of biopolymers with possible associated properties.

The biodegradable polymers gained tremendous interest of scientists and were developed using several novel technologies including tissue engineering, fabrication of responsive polymeric nanomaterials, edible food packaging, and additive manufacturing. This might be due to the great versatility of biodegradable polymers in terms of compatibility with other materials and additive processing approaches with customization of resulting devices. The researcher has investigated different classes of biodegradable polymers for additive manufacturing including proteins, polysaccharides, aliphatic polyesters, polyurethanes, *etc*. Additive manufacturing was defined by the American Society for Testing Materials (ASTM) in 2012 as the process of joining materials layer by layer that forms three-dimensional objects, controlled by computer-aided design and manufacturing software. Moreover, another breakthrough with biodegradable polymer is bioplastic innovation. Bioplastic has bought a significant revolution in the market with the potential to replace single-use synthetic petroleum product-derived plastic. In addition, valorization of the fruit husk from *Garcinia mangostana, Nephelium lappaceum, Durio zibethinus* [8], and *Tamarindus indica* [9] as biopolymers created another landmark in the fabrication of economic pharmaceutical products and biomedical devices.

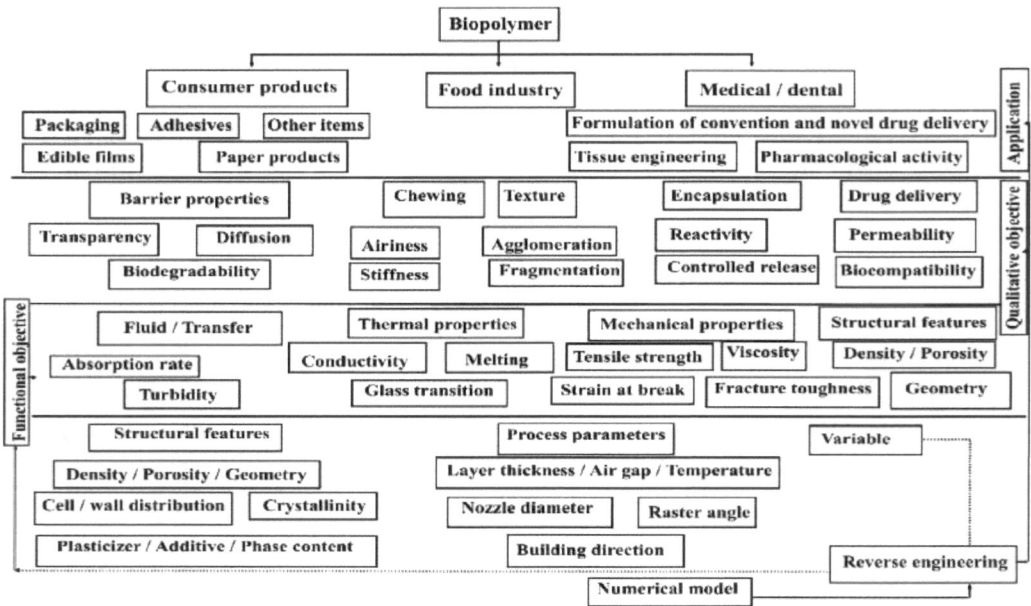

Fig. (3). Various applications of biopolymers with possible associated parameters and properties could affect the finished product.

MARKET POTENTIAL OF PHARMACEUTICAL NATURAL EXCIPIENTS

The comprehensive pharmaceutical excipient market showed moderate to exponential growth and increased amalgamation with the expansion of biodegradable adjuvants in the emerging market for several categories of products during the last five years. In 2015, the excipients manufacturing industry became one of the largest businesses with a 43% market capital share of total market value followed by alcohol such as propylene glycol of 20% and sugar 3% [10]. Although excipients play an important role in the pharmaceutical formulations by accumulating functionality within the product, however excipients' manufacturers need to respond to the varying pharmaceutical supply and demands considering their cost and other intermediates. Several new trends are emerging among pharmaceutical manufacturers including increased merging based on product-line enhancement or geographic expansion with selective investment in embryonic markets and targeted growth in selected product ranges. However, there are only a few manufacturers operating completely commercial scale-up production plants for the development of biodegradable polymers. This might be a possible reason that the market volume of biopolymers remains extremely low compared with petrochemical-originated synthetic polymers.

The global excipient market was valued at nearly $ 4.9 billion in 2011 according to the IMS Institute for Healthcare informatics and a market research firm Business Communications Company [10]. The valuation of excipients market by Business Communications Company reported considering several factors such as global pharmaceutical supply chain, application of quality by design by the manufacturer, overall drug safety concern, *etc*. Moreover, the international market of excipients was valued at 6.5% of the compound' annual growth rate (CAGR) in 2016. In addition, 11.1 billion pounds of excipients were consumed in 2011, increasing to 14 billion pounds by 2016 [10]. The demand for biodegradable polymers is driven by several factors including favorable regulation to reduce waste packaging and landfills, standard and certification procedure for packaging materials, composting infrastructures, and consumers' awareness of the re-organization of benefits from biodegradable polymers.

The overall excipients market is expected to grow from $ 8.3 billion in 2021 to $ 10.6 billion by 2026, with a CAGR of 5.0% for the period of 2021-2026 [11]. Similarly, the worldwide market for contract pharmaceutical manufacturing, research, and packaging that significantly contribute to the expansion of the excipients is the market expected to develop from $ 168.0 billion in 2021 to $ 214.7 billion by 2026, at a CAGR of 5.0% during the forecast period of 2021-2026 [12]. Whereas, micro packaging market is estimated to grow from $ 540.4 million in 2021 to $ 704.2 million by 2026 at a CAGR of 5.4% from 2021 to 2026 [13]. Moreover, with the overall growth of polymeric hydrocolloids [14] and water-soluble polymers technologies [15], the market is valued to reach $ 7.0 billion and $ 49.6 billion by 2022 at a CAGR of 4.9 and 5.8%, respectively for the period of 2017-2022. In addition, the market volume of biodegradable polymers is expected to propagate from $ 1.0 kilotons in 2021 to $ 1.9 kilotons by 2026 with a CAGR of 14.0% during 2021-2026 [16]. The market capital for guar gum [17] and xanthan gum [18] was valued at $ 659.55 and 699.0 million in 2016 and is expected to rise by 194.95% and 139.05% in USD, respectively by 2022. Furthermore, the estimated value of alginate in Latin America was $ 18.22 billion in 2018, which is expected to rise to $ 21.81 billion by 2023. The global consumption and distribution of sustainable and biodegradable polymers by geographical regions in 2018 are presented in Fig. (**4**) [19].

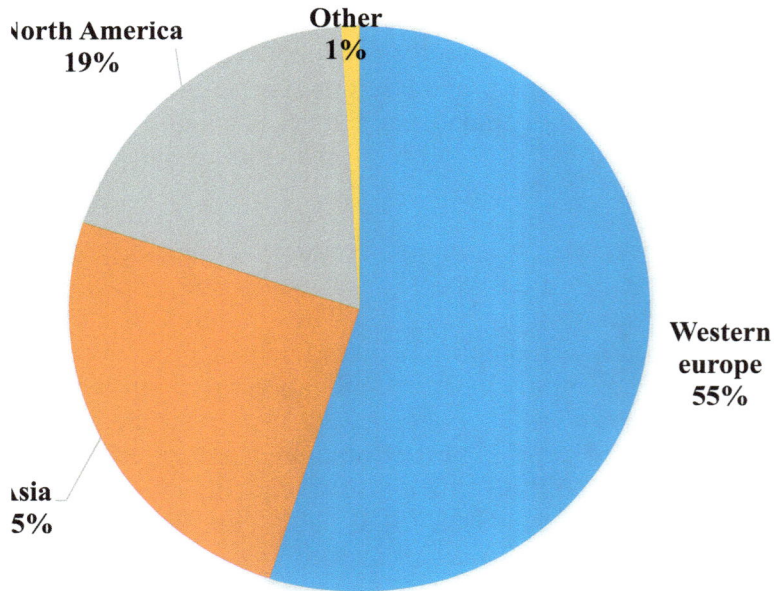

Fig. (4). Distribution of biodegradable polymer consumption worldwide as of 2018, by geographical region [28] (for interpretation of the results to color in this Fig. legend, the reader referred to either web version of this chapter or color print).

The recent market trends of biodegradable polymers show that a wide range of end-users are available even though potentially the market has yet not grown to its peak (Fig. **5**). Continued progress in terms of product development and cost reduction is required before biopolymers can effectively compete with conventional petrochemical-based synthetic polymers for typical applications. The major replacement for non-biodegradable plastics is indicated by the development of starch-based biodegradable plastics for the manufacturing of various types of bags, rigid packaging including thermoformed trays, containers, and loose-fill packaging foams. Similarly, starch-based products are used in the fabrication of agriculture, horticulture, and other household products such as mulching film, covering film, plant pots, cartridges, *etc*. However, the cost of some biodegradable plastics is still higher than synthetic polymer plastics. Thus, necessary awareness and education on the detrimental impact of non-biodegradable materials use can significantly boost the market potential of biodegradable polymers with a hike in its market share.

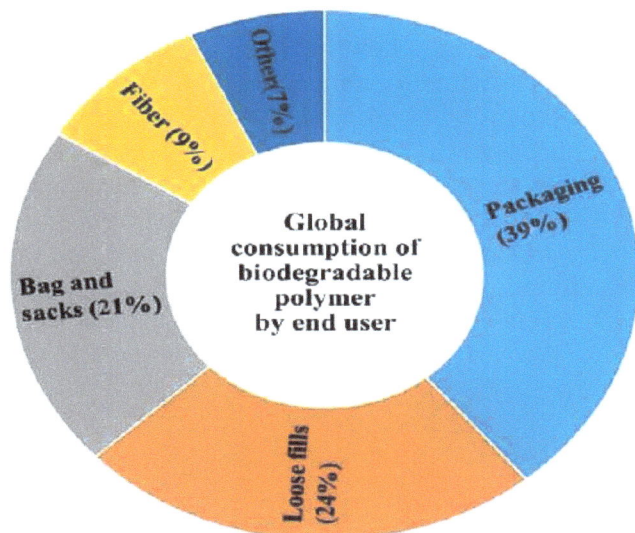

Fig. (5). Spectacles percentage share of global biodegradable polymer consumption by the end users [29] (for interpretation of the results to color in this Fig. legend, the reader referred to either web version of this chapter or color print).

BIOPOLYMERS LINKAGE WITH THE SUSTAINABLE AGRICULTURE INDUSTRY

Today the imagination of a world without synthetic polymers is very difficult, hitherto their large-scale fabrication and usage only date back to around 1950. Moreover, the growing world populations with several needs present various challenges, among them providing food and shelter to everyone, and the effect of non-biodegradable materials on the environment require an instant retort. Although, everyday science and technology create a new landmark and still the solutions to every problem are not found. Since 1950, a drastic upsurge has been observed in the usage of plastics and plastic-based products within the agriculture sector. The synthetic polymer-based materials for agriculture use were developed to provide durability and resistance to the fungal attack. By virtue of this anticipated market opportunity, several companies entered the arcade to manufacture a product using conventional resins similar to plastic with biodegradability. However, the manufacturing cost of the resins-based product was higher, compared to synthetic polymers. Later, another alternative to an expensive resin-based product was fabricated using polyethylene known as mulch. These mulches were manufactured using petroleum-based plastics, particularly polyethylene for agriculture use since it prevents crops and vegetables from weeds with control in soil temperature. Plastics mulching was first introduced to the agriculture sector in the 1950s and since then, it has successfully been used worldwide to increase productivity (Fig. **6**). However, major landfills were produced by polyethylene mulches that typically degrade by burning and produce airborne pollutants such as dioxane [20]. In addition, long-term use of mulches leads to embrittlement and subsequently fragmentation within the soil. Whereas, the bacteria that could break down plastic wastes were discovered in 1975 by Japanese researchers during the treatment of wastewater from the nylon industry with strains of Flavobacterium such as Flavaobasgteria and *Pseudomonas* [21]. The list of recently reported bacterial and fungal cultures involved in the degradation of pharmaceutical waste is presented in Table **1**.

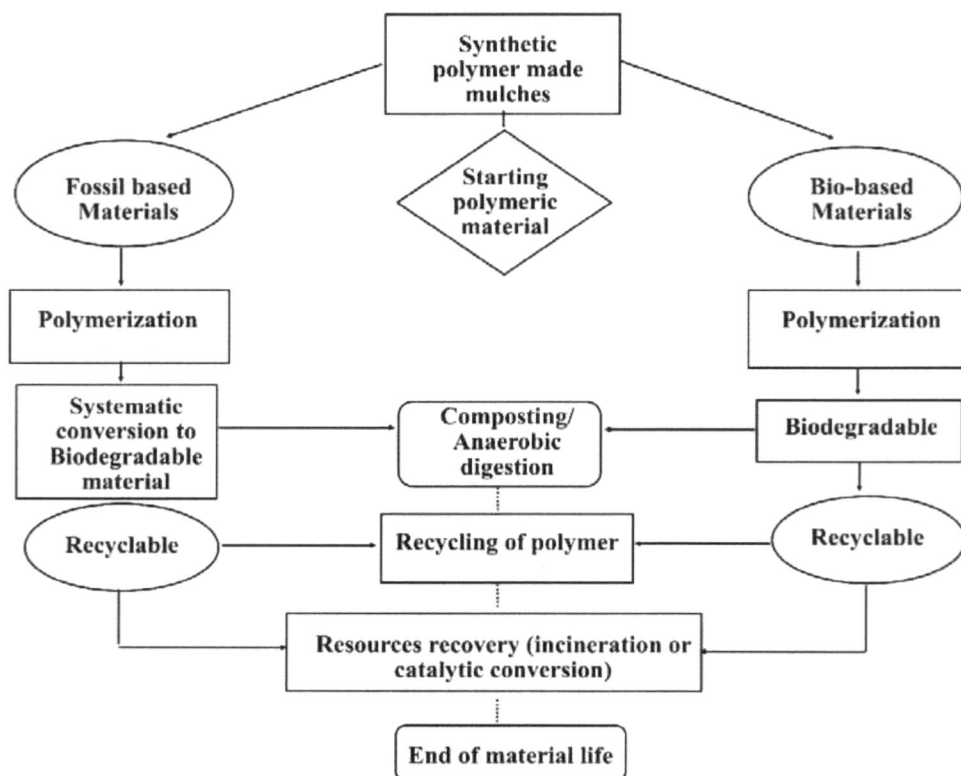

Fig. (6). Synthetic polymer based mulches and waste management strategies [4].

Table 1. Bacterial and fungal culture involved in pharmaceutical waste treatment and antibiotics degradation.

Bacterial Cultures	Waste Degradation	Reference
Arthrobacter, Citrobacter youngae, Enterobacter hormaechei, Pseudomonas sp., and *Rhodococcus* equi	Degradation of pharmaceutical mixture of NSAIDS	[22]
Castellaniella denitrificans	Degradation of sulfonamides antibiotics	[23]

(Table 1) cont.....

Bacterial Cultures	Waste Degradation	Reference
Flavobacterium johnsoniae, Pseudomonas aeruginosa, Pseudomonas pseudoalcaligenes, Paracoccus versutus, Moraxella osloensis, Sphingobacterium thalpophilum, and *Tsukamurella inchonensis*	Degradation of organic pollutants and bio-sorbent of toxic heavy metals	[24]
Candida inconspicua, Fusarium solani, Fusarium udam, Galactomyces pseudocandidum, and *Phaerochaete chrysosporium*	Degradation of organic pollutants	[24]
Pseudallescheria boydii, Rhodotorula mucilaginosa, Trichosporon asahi, Trichosporon domesticum	Degradation of organic pollutants	[24]

The global consumption of plastics for agricultural purposes was reported to be 4% with a volume of 2,850,000 tons in 2003 [25]. Perhaps the worldwide usage of plastics by farmers reduced to 2% in 2010 and reached 6.96 million tons in 2017, the volume of waste generated is difficult to dispose of [26]. Non-biodegradable polymeric materials for agriculture use were listed by PasticEurope including greenhouses, tunnels, mulching, plastic reservoirs, irrigation systems, crates for crop collection [27], *etc.* Fig. (7) shows the cumulative plastic waste generation and disposal in million metric tons.

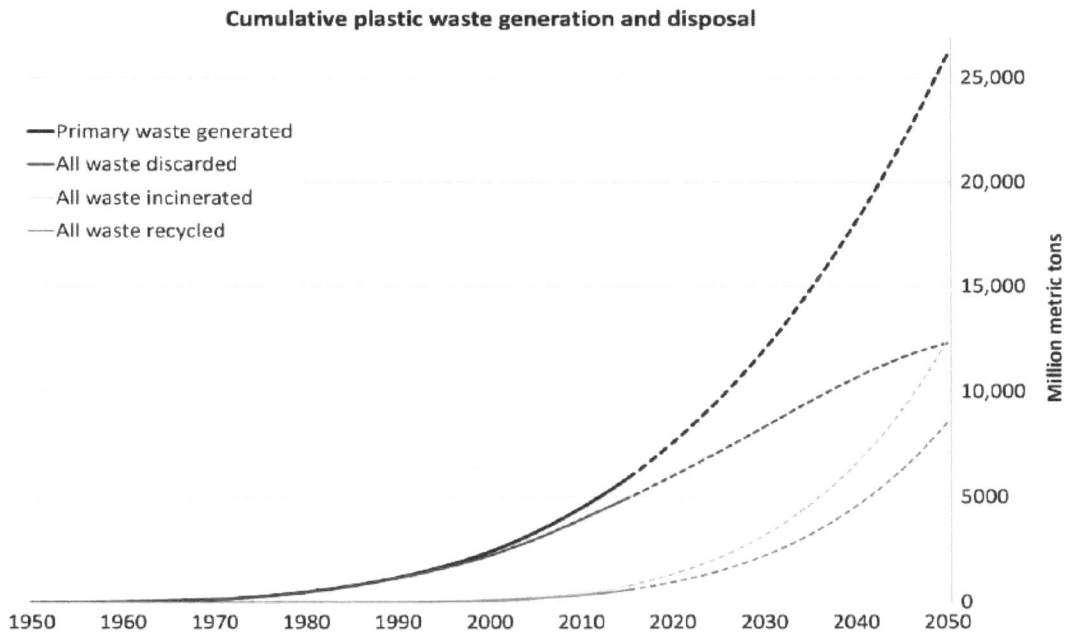

Fig. (7). Accumulative plastic waste spawned and disposed of in million metric tons until date with an expected approximate volume until 2050. Solid chronological lines show past data from 1950 to 2015, whereas dashed lines show forecasts of bygone trends [30].

FOOD AND DRUG ADMINISTRATIVE REGULATION

Adherence to food and drug administrative regulation is a key factor towards the assurance of safety and efficacy of adjuvants added within pharmaceuticals for medical use. However, non-compliance to the regulation and notwithstanding care and precision during manufacturing might ruin the quality as well as the efficacy of therapeutics with severe health issues. The major official agencies responsible for the standardization of excipients include the United States Food and Drug Administration, the European Union, Swiss medic, *etc.*, coordinated by the International Council for Harmonization of Technical Requirements for Pharmaceuticals for human use. In addition to official regulatory agencies, several other associations also guide on supporting documents and provide details on the regulations with their application in specific situations. Some of the most influential organizations relevant to the pharmaceuticals are the International Society for Pharmaceutical Engineering, the United States Pharmacopeia, Parenteral Drug Association, Indian Pharmacopeia, and the World Health Organization. The United States Food and Drug Administration states that any inactive ingredient incorporated within the pharmaceutical should be "Generally Recognized as Safe" or regulated as adjuvants and used within the recommended limits. Biopolymers and their products are generally considered safe that completely degrade with complete elimination from the body by a natural metabolic process without producing any side effects. However, the United States Food and Drug Administration approved some biodegradable polymers such as poly(lactic-co-glycolic acid), poly(glycolic acid), poly(lactic acid) for a medical application that eliminate from the body in the form of carbon dioxide. Several other biodegradable polymers approved by the United States Food and Drug Administration for use in the formulation of pharmaceutical products are listed in Table **2**.

Table 2. List of major biopolymers approved by the united states food and drug administration for incorporating with therapeutics in pharmaceutical formulations [31].

Biopolymer	Routes of Administration	Dosage Form	Maximum Potency Per Unit	Maximum Daily Exposure Limit
Acacia	Oral	Powder and suspension	648 – 1231 mg	7386 mg
Aspartame	Oral	Tablet and suspension	3.70 – 12 mg	14.8 – 18 mg
Carboxymethyl starch	Oral	Capsule and tablet	15 mg	-

(Table 2) cont.....

Biopolymer	Routes of Administration	Dosage Form	Maximum Potency Per Unit	Maximum Daily Exposure Limit
Carboxymethyl cellulose	Oral and intramuscular	Tablet, injection, capsule, drops, powder, suspension, jelly, film, dental enteral, ophthalmic, *etc.*	5 – 241 mg	18 – 347 mg
Carrageenan	Dental, nasal, oral, inhaler, and topical	Paste, powder, capsule, suspension, tablet, lotion, and film	0.42 – 20.15 mg	1 – 15 mg
Cellulose microcrystalline cellulose	Nasal and oral	Spray, capsule, granule, powder, suspension, and tablet	13.1 – 1975 mg	60 – 800 mg
Collagen	Topical and oral	Gel, lotion, shampoo, liquid, and capsule	75 – 270 mg	-
Dextran	Ophthalmic and intravenous	Solution, injection, and powder	300 mg	-
Dextrose, dextrose monohydrate	Oral, intramuscular, intra-spinal, intravenous, and nasal	Lozenge, injection, powder, solution, spray, capsule, granule, pastille, syrup, and troche	50 – 1600 mg	200 – 30144 mg
Galactose, galactose monohydrate	Oral and rectal	Powder, solution, and tablet	0.67 – 4.5 mg	-
Gelatin	Buccal, dental, intracavitary, intramuscular, intravenous, subcutaneous, nasal, oral, periodontal, and inhaler	Gum, paste, injection, powder, injection, solution, drops, capsule, elixir, and pastille	14 – 65 mg	288 – 4848 mg
Guar gum	Buccal, ophthalmic, oral, topical, and vagnial	Tablet, suspension, capsule, liquid, powder, lotion, and gum	0.12 – 35.4 mg	6 – 240 mg
Hypromellose	Oral, nasal, ophthalmic, rectal, and topical	Suspension, tablet, troche, gel, cream, jelly, lotion, shampoo, solution, capsule, and granule	1.7 – 221 mg	10 – 1117 mg
Karaya gum	Buccal	Tablet	68.1 mg	-
Lactose, lactose monohydrate	Buccal, intramuscular, intravenous, and oral	Tablet, injection, powder, solution, capsule, and granule	120 – 1682 mg	150 – 4384 mg
Lanolin	Topical, nasal, ophthalmic, and rectal	Spray, ointment, cream, lotion, and emulsion	1.5 – 3.0 (% w/w)	-

(Table 2) cont.....

Biopolymer	Routes of Administration	Dosage Form	Maximum Potency Per Unit	Maximum Daily Exposure Limit
Lecithin	Oral, rectal, topical, and vaginal	Injection, capsule, powder, suspension, tablet, suppository, gel, and cream	0.35 – 1 (% w/w)	1 – 3900 mg
Low-substituted Hydroxypropyl cellulose	Oral	Tablet and capsule	25 – 94.29 mg	13 – 480 mg
Mannitol	Intravenous, ophthalmic, oral, and topical	Injection, capsule, granule, powder, lozenge, suspension, tablet, torche, wafer, cream, and emulsion	37.11 – 750 mg	14 – 6000 mg
Methylcellulose	Buccal, intraarticular, intramuscular, nasal, ophthalmic, oral, sublingual, and topical	Cream, tablet, injection, jelly, drops, solution, capsule, powder, suspension, syrup, aerosol, and sponge	0.1% w/v	-
Microcrystalline cellulose	Oral, buccal, intravitreal	Tablet, capsule, granule, pellet, powder, torche	1.66 – 789.6	144 – 29520 mg
Pectin	Dental, nasal, and oral	Paste, spray, capsule, powder, and tablet	5.45 – 255 mg	~ 1 mg
Pullulan	Oral	Tablet	-	10 mg
Saccharin	Dental, oral, respiratory, sublingual, and topical	Paste, powder, drops, elixir, granule, solution, suspension, syrup, and tablet	0.9 – 20 mg	8 mg
Sodium alginate	Oral and topical	Capsule, lozenge, suspension, syrup, tablet, troche, and liquid	80 – 320 mg	700 – 955 mg
Sodium starch glycolate	Buccal, oral, and sublingual	Tablet, capsule, and powder	10 – 876 mg	5 – 540 mg
Starch	Buccal and oral	Tablet, capsule, granule, powder, suspension, wafer, suppository, cream, and inserts	0.17 – 430 mg	-
Sucrose	Buccal, intravenous, and oral	Lozenge, tablet, troche, injectable liposome, injectable emulsion, suspension, powder, capsule, pellets, drops, elixir, granule, lozenge, syrup, and tablet	109 – 900 mg	10 – 12000 mg

(Table 2) cont.....

Biopolymer	Routes of Administration	Dosage Form	Maximum Potency Per Unit	Maximum Daily Exposure Limit
Tragacanth	Buccal, oral, and topical	Tablet, powder, suspension, wafer, and jelly	5 – 60 mg	12 – 70 mg
Xanthan gum	Ophthalmic, oral, rectal, and topical	Solution, capsule, film, granule, liquid, lozenge, paste, powder, suspension, tablet, troche, enema, suspension, aerosol foam, cream, gel, lotion, chewing gum, film, and spray	0.01 – 275 mg	1 – 600 mg

CONCLUSIONS AND FUTURE PERSPECTIVE

Advancement in the human lifestyle and enormous utilization of non-biodegradable products has raised a serious concern about their surrounding and future impact. In addition, the consumption of polymers manufactured using synthetic materials engenders a severe impact on human beings as well as on the environment. To sustain life and maintain biological functions, nature requires selectively tailored molecular assemblies, and interfaces that provide specific chemical functions and structure, as well as a change in their environment. Although environmental sustainability is consistently gaining importance among individuals, however their utilization process towards green and sustainable development is moderate. In the meantime, it is not an easy way to completely avoid synthetic polymer and its product use due to its inherent quality, particularly in densely populated countries. Therefore, government and regulatory bodies should implement the legislation-governing increase in the production and use of biodegradable polymers for sustainable environment and economic development. Further, directive policies are required towards climate change and global warming due to the extensive use of non-degradable materials that produce toxic gas and initiate imbalance in the carbon cycle within the environment. Furthermore, recycling or degradation of existing non-biodegradable materials manufactured from synthetic polymers must be of prime importance.

REFERENCES

[1]   Pifferi G, Restani P. The safety of pharmaceutical excipients. Farmaco 2003; 58(8): 541-50.
      [http://dx.doi.org/10.1016/S0014-827X(03)00079-X] [PMID: 12875883]

[2]   Lazzarotto IP, Ferreira SD, Junges J, *et al.* The role of CaO in the steam gasification of plastic wastes recovered from the municipal solid waste in a fluidized bed reactor. Process Saf Environ Prot 2020; 140: 60-7.
      [http://dx.doi.org/10.1016/j.psep.2020.04.009]

[3]   Ugoeze K, Amogu E, Oluigbo K, Nwachukwu N. Environmental and public health impacts of plastic wastes due to healthcare and food products packages: a review. J Environ Sci Public Health 2021; 5: 1-31.

[4]   Amulya K, Katakojwala R, Ramakrishna S, Mohan SV. Low carbon biodegradable polymer matrices for sustainable future. Composites Part C: Open Access 2021.
[http://dx.doi.org/10.1016/j.jcomc.2021.100111]

[5]   Nutton V. Ancient medicine. 2nd ed. Routledge 2012; p. 504.
[http://dx.doi.org/10.4324/9780203081297]

[6]   Troy DB. The science and practice of pharmacy. In: Williams and Wilkins Lippincott, Ed. Philadelphia, PA 1974.

[7]   Vroman I, Tighzert L. Biodegradable polymers. Materials (Basel) 2009; 2(2): 307-44.
[http://dx.doi.org/10.3390/ma2020307]

[8]   Chansiripornchai P, Pongsamart S. Treatment of infected open wounds on two dogs using a film dressing of polysaccharide extracted from the hulls of durian (Durio zibethinus Murr.): Case report. Wetchasan Sattawaphaet 2008; 38(3): 55-61.

[9]   Zhongming Z, Linong L, Wangqiang Z, Wei L. Tamarind shells converted into an energy source for vehicles 2021. Available from: https://www.ntu.edu.sg/news/detail/converting-tamarind-shells-i-to-an-energy-source-for-vehicles

[10]   Bahadur S, Roy A, Chanda R, Choudhury A, Das S, Saha S, *et al.* Natural excipient development: need and future. Asian J Pharm Res 2014; 4(1): 12-5.

[11]   Tcherpakov M. Excipients in pharmaceuticals: global markets to 2026: business communications company research. 2021. Available from: https://www.bccresearch.com/market-research/phar-maceuticals/excipients-in-pharmaceuticals-global-markets.html

[12]   Staff BP. Global markets for contract pharmaceutical manufacturing, research and packaging: business communications company research. 2021. Available from: https://www.bccresearch.com/market-research/pharmaceuticals/global-markets-for-contract-pharmaceutical-manufacturing-resea-ch-and-packaging.html

[13]   Staff BP. Micropackaging: Global Markets to 2026: Business Communications Company Research.. 2021. Available from: https://www.bccresearch.com/market-research/advanced-materials/mic-o-packaging-market.html

[14]   Staff BP. Hydrocolloids: Technologies and Global Markets: Business Communications Company Research. 2018. Available from: https://www.bccresearch.com/market-research/advance--materials/hydrocolloids-technologies-and-global-markets-report.html

[15]   Sengupta A. Water-soluble Polymers: Technologies and Global Markets: Business Communications Company Research. 2017. Available from: https://www.bccresearch.com/market-research/advanced-materials/water-soluble-polymers-technologies-and-global-markets.html

[16]   Staff BP. Biodegradable polymers: global markets and technologies: business communications company research. 2021. Available from: https://www.bccresearch.com/market-research/plastics/biodegradable-polymers-global-markets-and-technologies.html

[17]   Research VM. Global guar gum market: business communications company research. 2018. Available from: https://www.bccresearch.com/partners/verified-market-research/global-guar-gum-market.html

[18]   Jeevane Y. Xanthan gum: applications and global markets: business communications company research. 2017. Available from: https://www.bccresearch.com/market-research/advance--materials/xanthan-gum-global-markets-report.html

[19]   Staff BP. Latin America: alginates market revenue 2018-2023: Business Communications Company Research. 2021. Available from: https://www.statista.com/statistics/983534/alginates-market-val-e-latin-america/

[20]   Hayes DG, Dharmalingam S, Wadsworth LC, Leonas KK, Miles C, Inglis DA. Biodegradable agricultural mulches derived from biopolymers. In: Kishan Khemani CS, Ed. Degradable Polymers and Materials: Principles and Practice. 2nd ed. Washington: American Chemical Society 2012; pp. 201-23.

[21]   Watanabe Y, Morita M, Hamada N, Tsujisaka Y. Formation of hydrogen peroxide by a polyvinyl alcohol degrading enzyme. Agric Biol Chem 1975; 39(12): 2447-8.

[22]   Aissaoui S, Ouled-Haddar H, Sifour M, Beggah C, Benhamada F. Biological removal of the mixed pharmaceuticals: diclofenac, ibuprofen, and sulfamethoxazole using a bacterial consortium. Iran J Biotechnol 2017; 15(2): 135-42.
[http://dx.doi.org/10.15171/ijb.1530] [PMID: 29845061]

[23]   Levine R. Microbial degradation of sulfonamide antibiotics: A Thesis. Springer 2016.

[24]   Rozitis D, Strade E. COD reduction ability of microorganisms isolated from highly loaded pharmaceutical wastewater pre-treatment process. J Mater Environ Sci 2015; 6: 507-12.

[25]   Jouet J. Situation da la plasticulture dans le monde. CIPA France 2003.

[26]   Picuno P. Innovative material and improved technical design for a sustainable exploitation of agricultural plastic film. Polym Plast Technol Eng 2014; 53(10): 1000-11.
[http://dx.doi.org/10 1080/03602559.2014.886056]

[27]   Group PMR. Plastics-The facts 2018. Plastics Europe. 2018. Available from: https://plasticseurope.org/wp-content/uploads/2021/10/2018-Plastics-the-facts.pdf

[28]   Tiseo I. Distribution of global biodegradable polymer consumption by region 2018: Statista. 2021. Available from: https://www.statista.com/statistics/891774/global-consumption-of-biodegrada-le-polymers-by-region/

[29]   Platt DK. Executive Summary. Biodegradable polymers: market report. UK: iSmithers Rapra Publishing 2006; pp. 5-10.

[30]   Geyer R, Jambeck JR, Law KL. Production, use, and fate of all plastics ever made. Sci Adv 2017; 3(7): e1700782.
[http://dx.doi.org/10.1126/sciadv.1700782] [PMID: 28776036]

[31]   Inactive Ingredient Search for Approved Drug Products United State of America: U.S. Food and Drug Adminstration. 2021. Available from: https://www.accessdata.fda.gov/scripts/cder/iig/index.cfm

Impact of Non-Biodegradable Polymers on the Environment and Human Health

Abstract: Synthetic polymers have been thriving in global industries over the past few decades due to their malleability, resilience, and economic value. But leaching of additives such as bisphenol-A, polybrominated diphenyl ether, and phthalates used in the manufacturing of polymeric products has raised serious concerns. However, the growing interest and investment in the development of biodegradable polymers could be a vital step toward reducing the impact of non-degradable polymers on the environment. Moreover, a combination of petroleum products with biopolymers can be a turning point for gradually replacing synthetic polymers to address or resolve these problems. In addition, a possible reduction in plastic polymer usage and manufacturing of products with materials that are less aggressive towards the environment can also reduce the impact of plastic on nature. Nature-derived biopolymers possess an enormous advantage over synthetic polymeric materials through cost-effectiveness, eco-, and user-friendly materials. Furthermore, the advanced applications of biopolymers in medical, tissue engineering, food industry, and fabrication of biotechnological products suggest that biopolymers are a boon for nature over synthetic polymers. This chapter discusses the advantage of biopolymers over synthetic polymers considering socioeconomic, human health, and environmental aspects. Additionally, the impact of petroleum-based polymeric materials on the environment compared to biodegradable polymers has been taken into consideration. The discussion is further extended to life cycle assessment, regulation, valorization, and utilization of polymer derived from waste with their potential use as inactive materials.

Keywords: Bio-based polymer, Environment, Life cycle assessment, Natural polymer, Valorization of waste.

INTRODUCTION

Living creatures produce a wide range of polymers as a significant part of their morphological, cellular, and dry matter. These biopolymers play a vital role in the life cycle of organisms to support their essential metabolic and cellular activities (Fig. **1**). Biopolymers are produced in the cytoplasm, organelles, cytoplasmic membrane, cell wall components, and the surface of cells even extracellularly by enzymatic processes. Today the life of human beings has become inconceivable without the use of polymeric products. Synthetic polymers are known to produce non-biodegradable products originating from the petroleum industry as monomers, which are subjected to specific chemical reactions including polymerization, poly-condensation, and poly-addition under a particular condition. Petrochemical plastics have many technical advantages that have significantly replaced other materials in many applications. These polymeric products can adopt different forms such as single-use plastics, bags, bottles, jars, thin films, multiple-use pipes, *etc*.

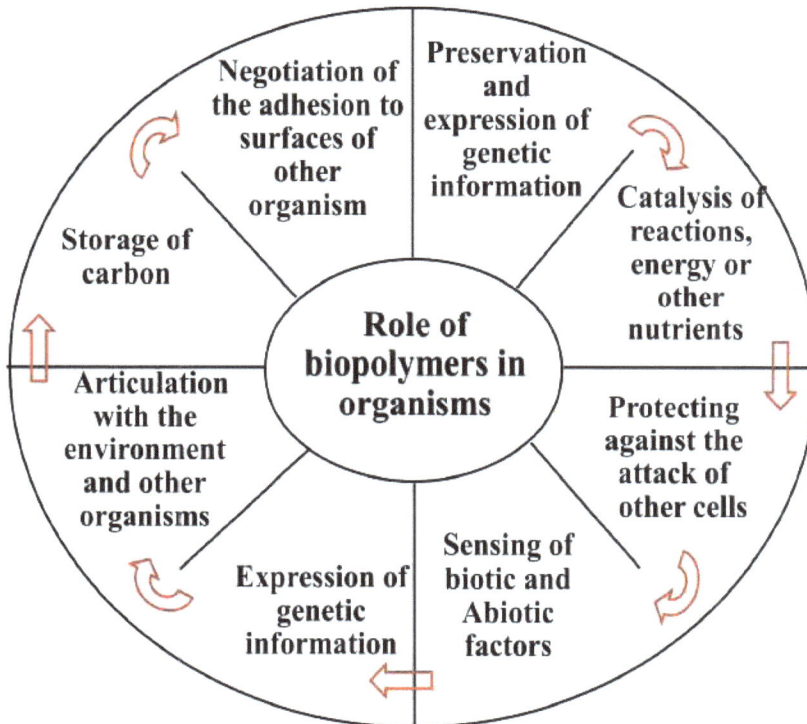

Fig (1). Biological role of biopolymers in a living organisms [1].

Moreover, synthetic polymers are also widely used in the fabrication of pharmaceuticals and medical products to regulate the life cycle. In addition, synthetic polymers and their products demonstrate good thermal and electrical insulators that resist corrosion and other associated chemical reactions with ease of handling. Although synthetic polymers can produce economic products, but their non-biodegradability and increasing accumulation in the environment every year cause detrimental effects on the ecosystem. Additionally, when synthetic polymer-based plastic enters the natural environment as litter through poor waste management or incorrect disposal, this poses a threat to wildlife too. Moreover, the leaching of additives such as bisphenol A, polybrominated diphenyl ether, and phthalates used in the manufacturing of polymeric products has also raised serious concerns about human health. To address these complications, researchers developed several biodegradable polymers originating from nature or chemically synthesized materials that can partially or completely replace the non-biodegradable polymeric product. Biopolymers have emerged as potential alternatives, some of which are available commercially while others remain under research. A biopolymer made from annually renewable resources which biodegrade can appear to solve the major problems associated with non-biodegradable polymers and their materials. In this chapter, an overview is presented on the impact of non-biodegradable polymers on the environment and their life cycle assessments. Further, a summary of recent research and review published on the biodegradation of natural polymers is provided. Furthermore, the effect of non-biodegradable polymers on human health has been reviewed.

Sustainability and Life-Cycle Assessments of Polymers on Eco-System

Biopolymers are generally considered an eco-friendly alternative to petrochemical polymers due to the renewable feedstock used to produce them and their biodegradability. Natural polymers are biosynthesized by living organisms *via* various techniques in the biosphere. Generally, such materials are high molecular weight polymers that break down to lower macromolecular structures upon the action of micro or macro-organism and enzymes [2]. In addition, catalysts promoting the degradation in nature *via* catabolic reactions followed by enzymes are categorized in six different classes depending on the catabolism [3]. The biodegradable polymeric materials including proteins, peptides, nucleic acids, lipids, natural rubber, lignin, and polysaccharides are obtained from various sources originating either from plants or microorganisms, however, the degradation rate differs depending on the nature of the functional group and the degree of complexity. Biopolymers are organized and classified variously as briefed in Chapter 1. The categorization of natural polymers allows the usage of starting monomers of diverse sequence and conformations at molecular, nano, micro, and macroscale, forming biodegradable multifarious polymers [4 - 6]. Conversely, synthetic biodegradable polymers degrade *via* oxidation, hydrolysis, thermal degradation, or other modes. The major advantage of biopolymers over synthetic polymers includes the potential to create a sustainable industry as well as enhancement in various properties such as durability, flexibility, high gloss, clarity, and tensile strength. However, several questions and challenges are also associated with biopolymers, since biopolymers are somehow produced using renewable raw materials that do not imply they are sustainable. Moreover, the farming practices used to grow agro-based polymers often carry significant environmental burdens, and the production energy can be higher than for petrochemical polymers [7]. Therefore, it is necessary to understand the potential environmental impacts of these new biopolymers developed either from natural resources or synthesized chemically *via* life cycle assessments.

The life cycle assessment is an environmental management tool for "collecting and evaluating the inputs, results, and potential environmental impacts of a system or product throughout its life starting from the extraction of raw materials and ending at the waste product being returned to the earth" [8]. Life cycle assessments involve collecting information on the inputs and outputs, such as emissions, waste, and resources of a process, and translating these to environmental consequences such as contributing to climate change, smog creation, eutrophication, acidification, and human and ecosystem toxicity [9]. Moreover, life cycle assessments support in identifying various opportunities to improve the environmental aspects of products at various points in their life cycle, in decision-making for both corporate and government organizations, and the selection of relevant environmental performance indicators with measurement and marketing techniques. The results obtained from the life cycle assessment are presented as damaged category followed by normalization. Normalization can be performed by dividing impact category scores by the average personal annual contribution allowing for the combination of categories [10]. The life cycle assessment distinguishes into four phases such as the definition of scope and objective, inventory analysis, impact assessments, and interpretation to reduce product impact on the eco-system [11]. Although, life cycle assessments have been standardized to a great extent by the International Organization for Standardization, however the guidelines leave much to be interpreted and selected by practitioners.

For petrochemical polymers, the life cycle starts with the extraction of raw materials required, including the fossil feedstock. The fossil feedstock is combined with non-renewable energy requirements to calculate the depletion of fossil fuels, usually presented as non-renewable energy use. This is possible by representing the feed-stock as energy rather than a materials input by multiplying the amount consumed by its heat of combustion [12]. Life cycle assessments of biopolymers derived from agro-industries additionally include data from the cultivation of the crop used. This includes the fuel required for farming activities such as plowing, agrochemicals applications, and harvesting. Further, it also includes the manufacture and transport of the materials required such as fertilizers, herbicides, and pesticides. Furthermore, land use, water consumption, and nitrogen emissions from fertilizers used may also be important factors. Other processes to include for typical biopolymers include milling and their produced products. Fig. (**2**) demonstrates research and reviews published based on life cycle assessment since 1987 to date with projected impact on the ecosystem in the tenure of non-degradable synthetic polymers imported from the Scopus database.

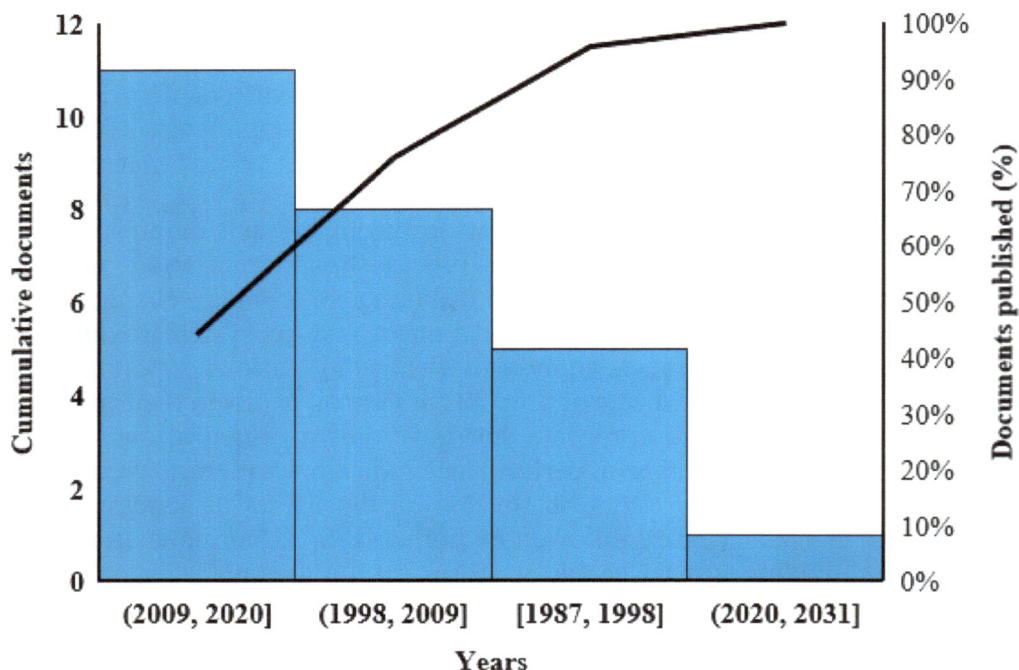

Fig. (2). Research and review published on the impact of non-biodegradable polymers on ecosystem available online on the Scopus database. The search includes Bullen terms such as "petroleum polymer*", "non-biodegradable*", "biopolymer*", "Impact*", "environment*" (Assessed from on 20rd Nov 2021) (For interpretation of the results to color in this Fig legend, the reader referred to either web version of this chapter or color print).

The Interactive Reaction Between Polymers and the Environment

The natural polymer degrades by a range of catabolic metabolisms catalyzed by a series of enzymes and produces energy that is utilized for the biosynthesis of new materials [13]. In addition, the cleavage of hydrolytically or enzymatically sensitive bonds in polymers leads to polymer erosion and biomaterial degradation [14]. Contrarily, the synthetic biodegradable polymers contain hydrolyzable linkages that are degraded by a hydrolytic process, such as ester, urea, and urethane linkage [15, 16]. The major bi-products produced by biodegradable polymers are carbon dioxide and water with some amount of ammonia gas. However, biologically stable non-biodegradable polymeric products including scaffolds, 3D-printed dressing, and orthopedic support systems can provide permanent support over the time course of application. Moreover, only a few biopolymers completely degrade to end up with carbon dioxide and water. Mostly, the degradation of biopolymers accumulate as metabolites and are used repeatedly in nature. Thus, it is very important to distinguish the type of smaller molecules/monomers formed during the degradation.

Biodegradable Oxidizable Polymers

The degradation product of oxidized polymer is generally complex in its pattern. Hydrolysis of a polymer yields one or at most three or four degradation bi-products. The hydrolysis of polysaccharides proceeds in three stages including a rapid decrease in molecular weight with slight changes in the weight (I), a decrease in the molecular weight change with extreme weight loss (II), and complete weight loss with the development of fragmented lactic acid and glycolic acids (III) [5].

The degradation of polymeric materials *via* oxidation is generally considered a slower process than hydrolysis. Several inert polymers and their polymeric materials degrade with slight modifications, for example, polyethylene undergoes reduction by photo-degradation or auto-oxidation *via* radical reaction [17]. In addition, the presence of low molecular weight monomers within polymeric products is more susceptible to biodegradation due to the ease of involvement with degradative chemical reactions. Moreover, oxidized polymers are brittle and hydrophilic in nature that degrades comparatively faster than a non-oxidized polymer. Gilead Scott complexed sulfur with dithiocarbamate for the first time to accelerate the embrittlement, which alone does not undergo thermal or photo-degradation [18, 19]. Thus, the incorporation of biopolymers with inert material might increase the susceptibility to biodegradation. Currently, several bio-based polymers including polysaccharides and starch are mixed together with transition metals to accelerate propensity towards photo-oxidation, thermolysis, and biodegradation. A combination of polyethylene with fillers such as granular starch indicated 10 times faster degradation than native polyethylene, which is an excellent example of abiotic reaction [20]. The mechanism behind this improvement in the degradation of polyethylene includes initial oxidation followed by removal of the two-carbon fragment from the polymeric chain, confirmed by the formation of carbonyl group on the polyethylene surface [21]. Several other compounds are also formed during the degradation of polyethylene depending on the type of reaction and degradation mechanism followed by assimilation of microorganisms. However, a recent report indicated that abiotic iron catalyzed through photo or thermos-oxidation is a rate-limiting step during bio-assimilation of photodegradable polyethylene [22]. The degradation mechanism and duration of biodegradable natural and synthetic polymers are tabulated in Table **1**.

Table 1. Degradative mechanism of biodegradable natural and synthetic polymers.

Biopolymer	Degradative Mechanism	Degradation (%)	Degradation Duration Reported (weeks)	Reference
Alginate	Enzymatic	>70	4	[40]
Gelatin	Hydrolysis and enzyme catalyzed decomposition	94.9	2.5	[41]
Chitosan	Enzymatic	60	4	[42]
Chitosan-gelatin	Enzymatic	28	4	[43]
		50-60	3	[44, 45]
Chitosan-polyvinylpyrrolidone-poly lactic-co-glycolic acid	Hydrolytic	27.1	4	[46]
Chitosan-silk fibroin	Enzymatic	50	4	[42]
Silk fibroin-hyaluronic acid	Enzymatic	47	3	[47]
Silk fibroin		72		
Collagen	Enzymatic	71	2	[48]
Collagen-poly-l-lactic acid	Hydrolysis and enzymatic	5		
Polylactic acid	Enzymatic	80	32	[49]
Polylactic acid-thermoplastic polyurethane	Hydrolytic	10	4	[50]
Polyurethane copolymer	Hydrolytic	10	4	[51]
Polyglycolide	Hydrolytic	50	1-6	[52]
Polycaprolactone	Hydrolytic	7	24	[53]
Poly lactic-co-glycolic acid	Hydrolytic	50	6	[54]
Polycaprolactone-poly-l-lactic acid	Hydrolytic	14	5	[54]
Starch-polyvinyl alcohol	Hydrolytic	27.1	4	[55]

Biodegradable Hydrolyzable Polymers

Recently considering the impact of polymer and polymeric products on the environment, several biopolymers have been developed as an alternative to traditional non-biodegradable synthetic polymers. In 1960, poly-L-lactide a natural, biocompatible, and biodegradable polymer with renewability *via* microbial fermentation was proposed for biomedical and pharmaceutical applications [23 - 26]. Furthermore, polycaprolactone, an aliphatic polyester, was discovered in the 1970s that can be easily hydrolyzed by microorganisms.

However, the degradation of polycaprolactone has been reported to be slower than other polymers such as poly(3-hydroxybutyrate) [27], poly[hydroxybutyrateco-valerate], cellulose, starch, and chitin [28 - 31]. Although biopolymer degrades easily and undergoes several modifications during processing or product development, therefore, a significant modification in biopolymer is required considering the functional property and biodegradability.

Effect of Non-Biodegradable Polymers on Wild Life and Human Health

Non-biodegradable polymers exhibit numerous societal benefits, offering future technological, and medical advances. However, concerns about the usage and disposal are diverse that include accumulation of waste in landfills and natural habitats, physical problems for humans and wildlife resulting from ingestion, leaching, and transfer of chemicals from the synthetic polymeric product, *etc*. The first accounts of plastic in the wildlife were reported from the carcasses of seabirds collected from shorelines in the early 1960 [32], the extent of the problem soon became unmistakable with plastic debris contaminating oceans from the poles to the equator and from shorelines to the deep sea. Most non-biodegradable polymers are buoyant in water, and utmost of plastic debris such as cartons and bottles often trap the air leading to substantial quantities of plastic debris accumulating on the sea surface that may also be washed ashore. Despite their buoyant nature, plastics can become fouled with marine life and sediment causing items to sink to the seabed. Moreover, plastic debris causes aesthetic problems, and it also presents a hazard to marine time activities including fishing and tourism. In addition, discarded floating plastic debris can rapidly become colonized by marine organisms resulting in interruption to non-narrative species. Several reports indicated that approximately 260 different species including invertebrates, turtles, fish, seabirds, and mammals ingested or entangled in plastic debris, resulting in impaired movement and feeding, reduced reproductive outputs, lacerations, ulcers, and death [33]. Additionally, the more serious concern was reported by the National Oceanic and Atmospheric Administration of the United States of America, indicating the presence of plastic fragments as small as 1.6 µm in some marine habitats. The experimental study suggested that small pieces are much more hazardous if consumed by filter feeders, detritivores, and or mussels that can retain in the body for at least 48 days and transfer the toxic substance to humans *via* seafood chain [34].

A range of chemicals known as additives used for the manufacturing of plastics are known to be toxic. Several reports suggested that this toxic chemical used potentially affects the human population including reproductive abnormalities [35]. Moreover, extensive evidence indicated that continuous exposure to plastics or their additives through ingestion, inhalation, or *via* dermal contacts increases the concentration of phthalates, bisphenol A (BPA), and polybrominated diphenyl ether with their metabolites in humans [36, 37]. In addition, there is some evidence of a negative association between phthalate metabolites and semen quality with an impaired level of testosterone [38]. Lang and coworkers reported a significant relationship between urine level and BPA with cardiovascular disease, type II diabetes, and abnormalities in liver enzymes [39]. However, the toxicological consequences of such exposure to children and pregnant women remain unclear and require further investigation. Despite the environmental concerns about some of the chemicals used in plastic manufacturing, it is important to emphasize that evidence for effects in humans is still limited and there is a need for further research and in particular, for longitudinal studies to examine temporal relationships with chemicals that leach out of plastics. Additionally, the traditional approach to study the toxicity of chemicals has been focused only on exposure to individual chemicals concerning disease or abnormalities. However, due to the complex integrated nature of the endocrine system, future studies involving endocrine-disrupting chemicals that leach from plastic products must focus on mixtures of chemicals to which people are exposed when they use common household products. Meanwhile, strategies are required to reduce the use of toxic additives (plasticizers) used in the manufacturing of plastics to control the effects on human populations. There is also a need for industry and independent researchers to work more closely with, rather than against, each other to focus effectively on the best way forward. The common heath issue associated with ingestion, inhalation, or *via* dermal contacts of synthetic non-biodegradable polymeric products is presented in Table **2**.

Table 2. A common source of phthalates and their health outcome [35].

Phthalates	Source of Exposure	Impact on Human Health
Dimethyl phthalate	Industrials petroleum production unit	Fetal toxicity
	Insect repellent	Affects musculosketal system, development of abnormalities of eye and ear
Diethyl phthalate	Personal care products such as shampoo, cosmetics, creams, soap	Reduces growth rate
	Pharmaceutical medication	Decreases food consumption
	Industrial solvent and pesticides	-

Phthalates	Source of Exposure	Impact on Human Health
Di-n-butyl phthalate	Adhesive	Decreases sperm production but returns to near normal levels after elimination of exposure
	Personal care products	Mild skin irritation
	Industrial solvent	-
	Medications	-
Butyl benzyl phthalate	Vinyl flooring	Testicular toxicity
	Adhesives and sealants	Cryptorchidism
	Synthetic leather	Reduces anogenital distance
	Car care products	Modulates steroids hormone levels
Di(2-ethylhexyl) phthalate	Soft plastics used in household products, tubing, toys, and floor tiles	Hepatocellular carcinoma, testicular toxicity
	Blood banks storage devices	Anovulation
	Food containers and packaging	Fetal growth restriction

CONCLUSIONS AND FUTURE PROSPECTS

Plastics offer considerable benefits for the future, but it is evident that our current approaches to production, use, and disposal are not sustainable and present serious concerns for wildlife and human health. The stringent regulation and serious milieu effect of conventional non-biodegradable polymers are driving forces for the development of renewable, biodegradable, sustainable, and ecofriendly materials. The recent trend indicates that polymer science and engineering research significantly shifted toward the development of biopolymers to reduce the impact of non-biodegradable polymers on ecosystems. In addition, ecologically supportive monetary policies and capital investment are continuously expanding in significance to people, condition, and business, among each other. Moreover, the development of noteworthy interest in the underdeveloped and developed country is an imperative step towards the reduction of plastic bioburden on the environment.

Since the production stage has been identified as the major contributor to the environmental burden of non-biodegradable polymeric materials and products, the use of renewable energy has been suggested as a future improvement, impacting nonrenewable energy and global warming potential. Another suggested improvement has been the use of alternative feedstocks that might be either waste-or by-products or less energy and chemical-intensive crops. Lastly, the development of enhanced degradable petrochemical polymer and improved framing practice for biopolymers productions could significantly reduce its impact on the environment and human health.

REFERENCES

[1]   Aggarwal J, Sharma S, Kamyab H, Kumar A. The realm of biopolymers and their usage: an overview. J Environ Treat Tech 2020; 8: 1005-16.

[2]   Barenberg SA, Brash JL, Narayan R, Redpath AE. Degradable materials. Taylor & Francis 2017.

[3]   Griffin GJ. Chemistry and technology of biodegradable polymers: Blackie Academic and Professional. 1994.

[4]   Council NR. Hierarchical structures in biology as a guide for new materials technology. National Academies Press 1994.

[5]   Karlsson S, Albertsson A. Biodegradable polymers and environmental interaction. Polym Eng Sci 1998; 38(8): 1251-3.
      [http://dx.doi.org/10.1002/pen.10294]

[6]   Hatada K. Macromolecular design of polymeric materials. Marcel Dekker 1997.

[7]   Yates MR, Barlow CY. Life cycle assessments of biodegradable, commercial biopolymers—A critical review. Resour Conserv Recycling 2013; 78: 54-66.
      [http://dx.doi.org/10.1016/j.resconrec.2013.06.010]

[8]   Finkbeiner M. The international standards as the constitution of life cycle assessment: the ISO 14040 series and its offspring Background and Future Prospects in Life Cycle Assessment. Springer 2014; pp. 85-106.

[9]   Rebitzer G, Ekvall T, Frischknecht R, *et al*. Life cycle assessment. Environ Int 2004; 30(5): 701-20.
      [http://dx.doi.org/10.1016/j.envint.2003.11.005] [PMID: 15051246]

[10] BS EN ISO-14044:2006 Environmental management–life cycle assessment–principles and framework: International Organization for Standardization. London: BSI 2006.

[11] Mercado G, Dominguez M, Herrera I, Melgoza RM. Are polymers toxic? Case study: environmental impact of a biopolymer. J Environ Sci Eng B 2017; 6: 121-6.

[12] (BSI) BSI. BS EN ISO-14040:2006 Environmental management life cycle assessment–principles and framework: International Organization for Standardization. London: BSI 2006.

[13] Wackett LP, Robinson SL. The ever-expanding limits of enzyme catalysis and biodegradation: polyaromatic, polychlorinated, polyfluorinated, and polymeric compounds. Biochem J 2020; 477(15): 2875-91.
      [http://dx.doi.org/10.1042/BCJ20190720] [PMID: 32797216]

[14] Nair LS, Laurencin CT. Biodegradable polymers as biomaterials. Prog Polym Sci 2007; 32(8-9): 762-98.
      [http://dx.doi.org/10.1016/j.progpolymsci.2007.05.017]

[15] Hsu S, Hung KC, Chen CW. Biodegradable polymer scaffolds. J Mater Chem B Mater Biol Med 2016; 4(47): 7493-505.
      [http://dx.doi.org/10.1039/C6TB02176J] [PMID: 32263807]

[16] Itävaara M, Siika-aho M, Viikari L. Enzymatic degradation of cellulose-based materials. J Polym Environ 1999; 7(2): 67-73.
      [http://dx.doi.org/10.1023/A:1021804216508]

[17] Gewert B, Plassmann MM, MacLeod M. Pathways for degradation of plastic polymers floating in the marine environment. Environ Sci Process Impacts 2015; 17(9): 1513-21.
      [http://dx.doi.org/10.1039/C5EM00207A] [PMID: 26216708]

[18]   Scott G. Developments in the photo-oxidation and photo-stabilisation of polymers. Polym Degrad Stabil 1985; 10(2): 97-125.
[http://dx.doi.org/10 1016/0141-3910(85)90024-2]

[19]   Al-Malaika S, Marogi AM, Scott G. Mechanisms of antioxidant action: Time-controlled photoantioxidants for polyethylene based on soluble iron compounds. J Appl Polym Sci 1986; 31(2): 685-98.
[http://dx.doi.org/10 1002/app.1986.070310232]

[20]   Albertsson AC, Barenstedt C, Karlsson S. Abiotic degradation products from enhanced environmentally degradable polyethylene. Acta Polym 1994; 45(2): 97-103.
[http://dx.doi.org/10 1002/actp.1994.010450207]

[21]   Albertsson AC, Barenstedt C, Karlsson S, Lindberg T. Degradation product pattern and morphology changes as means to differentiate abiotically and biotically aged degradable polyethylene. Polymer (Guildf) 1995; 36(16): 3075-83.
[http://dx.doi.org/10.1016/0032-3861(95)97868-G]

[22]   Arnaud R, Dabin P, Lemaire J, *et al.* Photooxidation and biodegradation of commercial photo-degradable polyethylenes. Polym Degrad Stabil 1994; 46(2): 211-24.
[http://dx.doi.org/10.1016/0141-3910(94)90053-1]

[23]   Kulkarni R, Pani K, Neuman C, Leonard F. Polylactic acid for surgical implants Walter Reed Army Medical Center Washington Dc Army Medical Biomechanical Ressearch Lab. Defence Technical Information Center 1966.

[24]   Kulkarni RK, Moore EG, Hegyeli AF, Leonard F. Biodegradable poly(lactic acid) polymers. J Biomed Mater Res 1971; 5(3): 169-81.
[http://dx.doi.org/10.1002/jbm.820050305] [PMID: 5560994]

[25]   Vert M, Li SM, Spenlehauer G, Guérin P. Bioresorbability and biocompatibility of aliphatic polyesters. J Mater Sci Mater Med 1992; 3(6): 432-46.
[http://dx.doi.org/10.1007/BF00701240]

[26]   Ghosh S. Recent research and development in synthetic polymer-based drug delivery systems. J Chem Res 2004; 2004(4): 241-6.
[http://dx.doi.org/10.3184/0308234041209158]

[27]   Nitkiewicz T, Wojnarowska M, Sołtysik M, *et al.* How sustainable are biopolymers? Findings from a life cycle assessment of polyhydroxyalkanoate production from rapeseed-oil derivatives. Sci Total Environ 2020; 749: 141279.
[http://dx.doi.org/10.1016/j.scitotenv.2020.141279] [PMID: 32818854]

[28]   Potts JECR, Ackart WB, Niegisch WD. The biodegradability of synthetic polymers. G J. Polymers and ecological problems polymer science and technology. Boston, MA: Springer 1973; pp. 61-79.
[http://dx.doi.org/10.1007/978-1-4684-0871-3_4]

[29]   Fields RD, Rodriguez F, Finn RK. Microbial degradation of polyesters: Polycaprolactone degraded by P. pullulans. J Appl Polym Sci 1974; 18(12): 3571-9.
[http://dx.doi.org/10.1002/app.1974.070181207]

[30]   Tokiwa Y. Degradation of polycaprolactone by a fungus. Food and Agriculture Organization of the United Nations 1976.

[31]   Tokiwa Y, Suzuki T. Hydrolysis of polyesters by lipases. Nature 1977; 270(5632): 76-8.
[http://dx.doi.org/10.1038/270076a0] [PMID: 927523]

[32]   Harper P, Fowler J. Plastic pellets in New Zealand storm-killed prions (Pachyptila spp.). Notorn Bird New Zealand 1987; 34(1): 65-70.

[33]   Gregory MR. Environmental implications of plastic debris in marine settings—entanglement, ingestion, smothering, hangers-on, hitch-hiking and alien invasions. Philos Trans R Soc Lond B Biol

Sci 2009; 364(1526): 2013-25.
[http://dx.doi.org/10.1098/rstb.2008.0265] [PMID: 19528053]

[34]   Browne MA, Dissanayake A, Galloway TS, Lowe DM, Thompson RC. Ingested microscopic plastic translocates to the circulatory system of the mussel, Mytilus edulis (L). Environ Sci Technol 2008; 42(13): 5026-31.
[http://dx.doi.org/10.1021/es800249a] [PMID: 18678044]

[35]   Kumar P. Role of plastics on human health. Indian J Pediatr 2018; 85(5): 384-9.
[http://dx.doi.org/10.1007/s12098-017-2595-7] [PMID: 29359236]

[36]   Crain DA, Eriksen M, Iguchi T, *et al.* An ecological assessment of bisphenol-A: Evidence from comparative biology. Reprod Toxicol 2007; 24(2): 225-39.
[http://dx.doi.org/10.1016/j.reprotox.2007.05.008] [PMID: 17604601]

[37]   Lyche JL, Gutleb AC, Bergman Å, *et al.* Reproductive and developmental toxicity of phthalates. J Toxicol Environ Health B Crit Rev 2009; 12(4): 225-49.
[http://dx.doi.org/10.1080/10937400903094091] [PMID: 20183522]

[38]   Pan Y, Jing J, Dong F, *et al.* Association between phthalate metabolites and biomarkers of reproductive function in 1066 Chinese men of reproductive age. J Hazard Mater 2015; 300: 729-36.
[http://dx.doi.org/10.1016/j.jhazmat.2015.08.011] [PMID: 26296076]

[39]   Lang IA, Galloway TS, Scarlett A, *et al.* Association of urinary bisphenol A concentration with medical disorders and laboratory abnormalities in adults. JAMA 2008; 300(11): 1303-10.
[http://dx.doi.org/10.1001/jama.300.11.1303] [PMID: 18799442]

[40]   Bahrami N, Farzin A, Bayat F, *et al.* Optimization of 3d alginate scaffold properties with interconnected porosity using freeze-drying method for cartilage tissue engineering application. Arch Neurosci 2019; 6(4): e85122.
[http://dx.doi.org/10.5812/ans.85122]

[41]   Wu X, Liu Y, Li X, *et al.* Preparation of aligned porous gelatin scaffolds by unidirectional freeze-drying method. Acta Biomater 2010; 6(3): 1167-77.
[http://dx.doi.org/10.1016/j.actbio.2009.08.041] [PMID: 19733699]

[42]   Bhardwaj N, Kundu SC. Silk fibroin protein and chitosan polyelectrolyte complex porous scaffolds for tissue engineering applications. Carbohydr Polym 2011; 85(2): 325-33.
[http://dx.doi.org/10.1016/j.carbpol.2011.02.027]

[43]   Hollister SJ, Maddox RD, Taboas JM. Optimal design and fabrication of scaffolds to mimic tissue properties and satisfy biological constraints. Biomaterials 2002; 23(20): 4095-103.
[http://dx.doi.org/10.1016/S0142-9612(02)00148-5] [PMID: 12182311]

[44]   Nieto-Suárez M, López-Quintela MA, Lazzari M. Preparation and characterization of crosslinked chitosan/gelatin scaffolds by ice segregation induced self-assembly. Carbohydr Polym 2016; 141: 175-83.
[http://dx.doi.org/10.1016/j.carbpol.2015.12.064] [PMID: 26877010]

[45]   Baniasadi H. Fabrication and characterization of conductive chitosan/gelatin-based scaffolds for nerve tissue engineering. International journal of Biolog Macromol 2015; 74: 360-6.

[46]   Saeedi Garakani S, Davachi SM, Bagher Z, *et al.* Fabrication of chitosan/polyvinylpyrrolidone hydrogel scaffolds containing PLGA microparticles loaded with dexamethasone for biomedical applications. Int J Biol Macromol 2020; 164: 356-70.
[http://dx.doi.org/10.1016/j.ijbiomac.2020.07.138] [PMID: 32682976]

[47]   Guan Y, You H, Cai J, Zhang Q, Yan S, You R. Physically crosslinked silk fibroin/hyaluronic acid scaffolds. Carbohydr Polym 2020; 239: 116232.
[http://dx.doi.org/10.1016/j.carbpol.2020.116232] [PMID: 32414432]

[48]   Xu C, Lu W, Bian S, Liang J, Fan Y, Zhang X. Porous collagen scaffold reinforced with surfaced activated PLLA nanoparticles. Sci World J 2012.

[http://dx.doi.org/10.1100/2012/695137]

[49]   Guo Z, Yang C, Zhou Z, Chen S, Li F. Characterization of biodegradable poly(lactic acid) porous scaffolds prepared using selective enzymatic degradation for tissue engineering. RSC Advances 2017; 7(54): 34063-70.
[http://dx.doi.org/10.1039/C7RA03574H]

[50]   Mondal S, Martin D. Hydrolytic degradation of segmented polyurethane copolymers for biomedical applications. Polym Degrad Stabil 2012; 97(8): 1553-61.
[http://dx.doi.org/10.1016/j.polymdegradstab.2012.04.008]

[51]   Jing X, Mi HY, Salick MR, *et al.* Morphology, mechanical properties, and shape memory effects of poly(lactic acid)/ thermoplastic polyurethane blend scaffolds prepared by thermally induced phase separation. J Cell Plast 2014; 50(4): 361-79.
[http://dx.doi.org/10.1177/0021955X14525959]

[52]   Shih YC. Effects of degradation on mechanical properties of tissue-engineering poly (glycolic acid) scaffolds. Yale Medicine Thesis Digital Library . 2015.

[53]   Lam CXF, Hutmacher DW, Schantz JT, Woodruff MA, Teoh SH. Evaluation of polycaprolactone scaffold degradation for 6 months *in vitro* and *in vivo*. J Biomed Mater Res A 2009; 90A(3): 906-19.
[http://dx.doi.org/10.1002/jbm.a.32052] [PMID: 18646204]

[54]   You Y, Min BM, Lee SJ, Lee TS, Park WH. *In vitro* degradation behavior of electrospun polyglycolide, polylactide, and poly(lactide-co-glycolide). J Appl Polym Sci 2005; 95(2): 193-200.
[http://dx.doi.org/10.1002/app.21116]

[55]   Mirab F, Eslamian M, Bagheri R. Fabrication and characterization of a starch-based nanocomposite scaffold with highly porous and gradient structure for bone tissue engineering. Biomed Phys Eng Express 2018; 4(5): 055021.
[http://dx.doi.org/10.1088/2057-1976/aad74a]

CHAPTER 3

Potential Sources of Biodegradable Polymers

Abstract: Synthetic polymers are an important class of pharmaceutical excipients that contribute significantly to the fabrication of different dosage forms. However, due to biodegradability concerns, the highly publicized disposal problem of traditional oil-based thermoplastics with a detrimental effect on the environment, has promoted the search for alternative biodegradable polymers. Biodegradable polymers are an ecofriendly, economic, and safe alternative to synthetic polymers due to their biodegradable nature and the source of origin. Biopolymers and biomaterials are available in abundance with different pharmaceutical and medical applications including drug delivery, wound healing, tissue engineering, imaging agents, *etc.* Moreover, biopolymers possess certain specific properties such as biocompatibility, biodegradability, low antigenicity, functionality to support cell growth, and proliferation with appropriate mechanical strength. Biopolymers are obtained from sustainable natural resources and animal processing co-products and wastes. Polysaccharides such as cellulose and starch represent the major characteristics of the family of these natural biopolymers, while other biodegradable polymers such as bacterial cellulose and sericin are also used to develop biodegradable materials. Recent advancements and development in the field of natural polymers have opened up new possibilities for the rational engineering of natural gums and mucilage towards the expansion of functional excipients suitable for industrial and medical applications. This chapter highlights the potential sources of novel biodegradable polymers with recent expansion in the processing of different novel natural polymers to develop multifunctional excipients and valorization of waste biomass to produce biopolymers.

Keywords: Biodegradable polymers, Biopolymers, Carrageen, Lignin, Sericin, Source of polymers.

INTRODUCTION

Biodegradable functional polymers were introduced in the 1980s and since then, these have been applied in the design and development of various components including the pharmaceutical dosage form, packaging, agriculture, and in several other allied areas [1]. Biodegradable polymers have received attention in the last decades due to their potential application in the maintenance of physical health with low cost of production, environmental protection, and delivery of active drug moiety. Biodegradable polymers are broadly distinguished into two classes such as synthetic and natural excipients. Natural polymers are those that are present in or created by living organisms. Moreover, such polymers are available in abundant from renewable sources, while synthetics are developed from non-renewable petroleum resources. The future of biodegradable polymer depends not only on the cost of production but also on its effectiveness in delivering the active component in a controlled manner. The term natural polymers refers to macromolecules that occur naturally or are synthesized by a living organism [2], whereas synthetic polymers are fabricated macromolecules. The synthesis of natural polymers generally involves enzyme-catalyzed, chain-growth polymerization reactions of activated monomers that are typically formed within cells by complex metabolic processes. Moreover, the degradation of natural polymeric materials generally occurs through enzymatic actions or *via* chemical deterioration. The enzymatic degradations include the breakdown of polymers in lower molecular mass through abiotic reactions, while hydrolysis and oxidation are chemical degradation pathways as presented in Fig. (**1**). Furthermore, recent discoveries in the field of bacterial polymer biosynthesis have opened up other new avenues for the rational and medical applications of biodegradable polymers. Generally, petroleum-derived polymers resist degradation *via* oxidation or hydrolysis, however, the addition of additives including antioxidant or pro-oxidants can initiate the biodegradation process. Fortification of additives within such polymer chain initiates the biodegradation process in the presence of ultraviolet light *via* photo-oxidation, which propagates by a free radical chain reaction in which the first hydrogen peroxide is produced and later a hydrophilic low molecular mass is formed by pyrolysis [3].

The evolution process of biodegradable polymers is yet not complete, or commercially successful. However, biodegradable polymers gained significant attention among researchers considering the utilization of renewable resources to replace declining and increasingly expensive fossil resources as a source for monomers and polymers. The selection of a suitable composition substrate is an important factor for the optimization of biodegradable excipients production and characterization. Since the high cost of substrates is the major contributing factor in biopolymers development and production cost, thus the usage of waste and biomass could significantly reduce the total fabrication cost. In this chapter, a detailed overview is presented on recent advances in the sources of biodegradable functional polymers either from natural or from synthetic manufacturing, with an outline on various emerging processed techniques utilized for the improvement of pharmaceutical properties as *via*ble multifunctional commercial replacements for currently used polymers. These biodegradable polymers were selected based on the ease of availability, the extent of knowledge available to describe their formation, and their commercial relevance with applied potential.

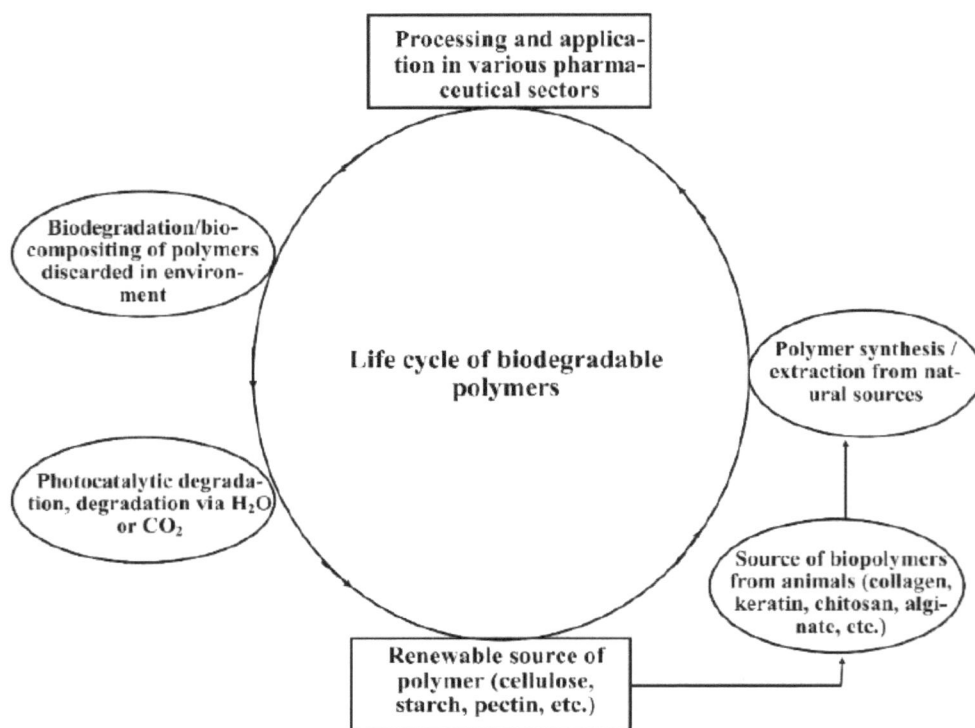

Fig. (1). The life cycle for biodegradable polymers obtained from natural resources.

Potential Polymers from Natural Resources

The source of excipients is often determined by nature and common practice with its potential applications. Depending on the source, traditionally excipients are distinguished as natural or synthetic. There are two main types of natural polymers (I) those that come from a living organism (carbohydrates and proteins) and (II) those that need to be polymerized and obtained from renewable resources (lactic acid and triglycerides). However, over the past few decades, a new class of multifunctional polymers resulted from the processing of natural and biodegradable synthetic polymers that have attracted significant pharmaceutical manufacturers and researchers. Some examples of such natural polymers are proteins and nucleic acid that occur in the human body, cellulose, natural rubber, silk, and wool. Several types of cellulose including starch and other carbohydrate-based natural polymers are made up of numerous glucose molecules linkage. Similarly, peptides, polypeptides, lignin, and proteins obtained from plants or animals have also been referred to as biodegradable natural polymers. Moreover, processed or grafted polymers are another class of naturally occurring excipients that are significantly used in the day-to-day life of human beings.

Silkworm Cocoon as a Potential Source of Biodegradable Polymers

Silk production includes a broad range of primarily protein-based polymers originated from insects (*Actias luna*, *Actias selene*, *Antheraea frithi*, *Antheraea mylitta*, *Antheraea perni*, *Antheraea polyphemus*, *Antheraea roylei cocoon*, *Antheraea roylei* shell, *Antheraea yamamai*, *Attacus atlas*, *Argema mimosae*, *Caligula cachara*, *Callosamia promethea*, *Cricula anderi*, *Cricula trifenestrata*, *Caligula simla*, *Hyalophora gloveri*, *Loepa katinka*, *Gonometa postica*, *Samia canningi*, *Samia cynthia*, *Opodiphthera eucalypti*, *Saturnina pavonia*, and *Saturnia pyri*), weaving spiders (*Nephila* clavipes), and silkworm (*Bombyx mandarina* and *Bombyx mori*) [4]. Silk is an excellent biomaterial that has been utilized in the textile industry. However, silk-made materials gained significant attention in the last few decades in the fabrication of biodegradable medical scaffolds, implantable devices, tissue engineering products, and additive manufacturing [5]. Cocoon shells are produced by silkworm caterpillars, a kind of natural polymeric material that possesses excellent mechanical properties. Production of cocoon shells is a biological process that involves the digestion of mulberry leaves to develop protective cocoons for pupas. This cocoon protects the moth pupa against potential predators and from microbial degradation or desiccation during metabolic activity [6, 7]. The process of cocoon formation by silkworm caterpillar involves cyclic spinning of a lightweight and compact cocoon in the shape of either "*8*" or "*S*" by blending and stretching its own body for approximately 3 days [8]. After, the completion of spinning of the cocoon, the silkworm sheds its skin to form a pupa. The silk cocoon shell is comfortable and protective, which allows the pupa in it to evolve into a silkworm moth. Depending on silkworm species and rearing environment, cocoons might vary in weight, color, stiffness, and thickness. The minimum thickness at the ellipsoidal end within the cocoon supports conversion of pupa to moth during the metamorphosis stage. Moreover, the non-uniform shrinking during drying causes the development of wrinkles on the outer surface of the cocoon [9]. In addition, the anti-parallel β-sheet structure forming microfibrils within silk fibers is responsible for its crystalline nature. These microfibrils are organized into fibril bundles to form a single continuous silk strand with a length in the range of 1000-1500 m and conglutinated by silk gum. The raw silk cocoons comprise basic two types of proteins, such as sericin, which is soluble in hot water, and insoluble fibroin [10]. The sericin is a group of glue proteins produced in the middle silk gland of silkworm that surrounds fibroin fibers and fixes them in the cocoons. Sericin contributes 20-30% of total cocoon weight, comprising 18 amino acids, including essential amino acids. The molecular heterogeneity study of sericin reported three different types of sericin protein [11], while at least six to fifteen with molecular weight ranged between 134-309kDa, reported using sodium dosulfatelphate–polyacrylamide gel electrophoresis analysis [12 - 14]. Whereas, fibroin is a natural amphiphilic amorphous or crystalline glycoprotein composed of two equimolar protein subunits of 370 and 25 kDa linked covalently by a disulfide bond. The presence of a dense chain of amino acids represents an amorphous form, while a high percentage of alanine, glycine, and serine indicates the crystalline form of fibroin. Moreover, a crystalline form (silk-I) and β-sheet form (silk-II) have been reported with molecular masses of 350-415K for fibroin [15]. The biological functions of silkworm cocoons including defense mechanism, thermal regulation, antioxidant, and antibacterial activity attracted researchers to understand the structural and morphological properties, to design and develop next-generation biodegradable polymers that can effectively bio-mimic the process of thermal regulations (Fig. 2).

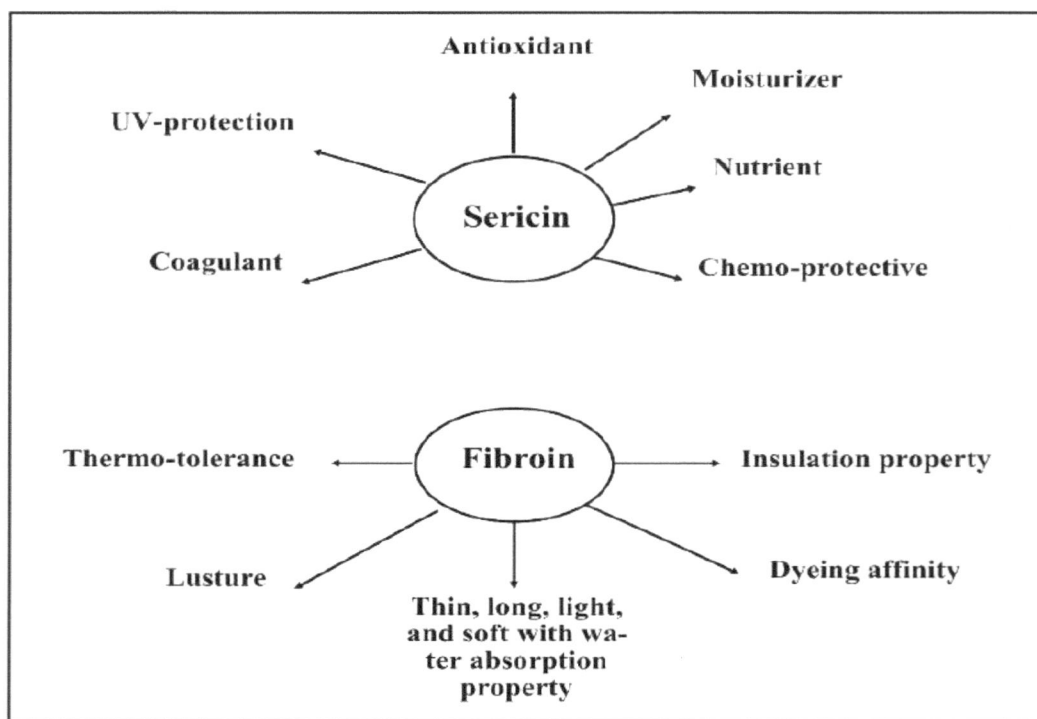

Fig. (2). Pharmaceutical, cosmeceutical, and textile attributes of silk and fibroin.

The Sericulture industry reported India as the second-largest producer of silk that produces nearly 16,700 mt silk/y with a price range of ₹ 900-1300/kg [16]. However, during processing, a huge amount of waste silk is generated, and sold at a very low price of ₹ 80-100/kg, which contributes 30% of the total cocoon production [17]. Moreover, silk waste is used in various sectors including fortified natural rubber with coarse yarn and spun silk, pharmaceuticals, and stabilization of metallic nanoparticles. As reflected from recent publication, fibroin and sericin have gained significant attention for the development of biomedical devices, biomaterials for wound healing, cosmetics, and antibacterial application that are presented in Table **1** and Fig. (**3**).

Table 1. Biomedical applications of silk sericin are obtained from different sources and process of extraction.

Biopolymer	Application	Reference
Silkworm cocoon	Flat silk mats were compared to cocoon derived silk mats and evaluated for mechanical, structural, and bone regeneration analysis.	[18]
Silk sericin	Conjugated alginate, sericin, and polyvinyl alcohol bio-absorbent demonstrated effective removal and recovery of ytterbium from wastewater.	[19]
Silkworm cocoon	Biomimetic and osteogenic three-dimensional composite scaffold with hydroxyapatite and nano magnesium oxide embedded in the fiber of silkworm and silk fibroin showed efficient bone repair potential.	[20]
Silkworm cocoon (***Antheraea assamensis***)	Muga silk obtained from *Antheraea assamensis* characterized structurally and results corroborated with the silk obtained from *Bombyx mori*.	[21]
Silkworm cocoon	Extraction process dependent antioxidant efficacy reported.	[22]
Silkworm cocoon	Silkworm-based silk fiber incorporated polyvinyl alcohol fibers demonstrated potential application as a new textile fabric.	[23]
Silk sericin	Silk sericin capped gold nanoparticles with antibacterial efficacy could be suitable nanoscale vehicles for biomedical application.	[24]
Silk sericin	*In situ* reduction of metallic nanoparticles within sericin/polyvinyl alcohol hydrogel-based wound dressing could facilitate re-epithelialization and collagen deposition to accelerate wound healing.	[25]
Silkworm cocoon (***Philosamia ricini***)	Silk sericin loaded curcumin nanoparticles demonstrated encapsulation efficacy in a range of 39 – 85% and results indicated that sericin isolated could be a potential polymer for nanoparticles fabrication.	[26]
Silkworm cocoon	Silk fibroin peptide demonstrated dose-dependent down-regulation of protein (Bcl-2) expression and up-regulation of bax protein in lung tumor cells (H460) during cell apoptosis.	[27]
Silkworm cocoon	Structural morphology and mechanical property of reared silkworm cocoon were tested and results indicated that the mechanical property was significantly different in three sections of the tested cocoon.	[28]
Silk sericin	Repetitive freeze-thawed silver nanoparticles fortified within agar exhibited an excellent antibacterial capability against potential wound pathogens.	[29]
Silk sericin	Silver nanoparticles incorporated within silkworm cocoon-based wound film could shed new light on the design of antibacterial biodegradable materials.	[30]

(Table 1) cont.....

Biopolymer	Application	Reference
Silkworm cocoon	*In situ* reduced silver nanoparticles in the presence of light incorporated within sericin composite showed potential for biomedical application.	[31]
Silkworm cocoon	Silkworm cocoon sol-gel wound dressing accelerated wound healing efficacy.	[32]
Silkworm cocoon	Silk protein demonstrated pseudoplastic behavior with internal strength when tested for antioxidant and emulsifier efficacy.	[33]
Silkworm cocoon	Both domestic and wild silkworm-based sericin incorporated composite tested for mechanical properties. Domestic cocoon indicated brittle and weak composite, while tough and strong strength observed for composite fabricated using wild cocoon based sericin.	[34]
Silkworm cocoon	Cell culture containing sericin indicated excellent substitute medium cell line study, compared to fetal bovine serum.	[35]
Silkworm cocoon	Silk protein demonstrated strong inhibitory activity against fungal protease.	[36]
Silkworm cocoon	Sodium alginate and sericin blend incorporated diclofenac sodium complexed with calcium chloride particles indicated particles size in the range of 91.0 – 75.0%.	[37]
Silkworm cocoon	Effect of carboxymethyl cellulose molecular weight on blended sericin hydrogel demonstrated enhancement in cell migration rate.	[38]
Silkworm cocoon	Silk sericin ameliorate the healing potential of burn wounds	[39]
Silkworm cocoon *Antheraea mylitta*	A lower molecular weight of silk protein was isolated from *Antheraea mylitta* cocoons.	[40]
Silkworm cocoon	Silk sericin protein demonstrated potential towards the health benefits and dietary supplements for colon cancer prevention.	[41]
Silkworm cocoon	Regulated swelling and release of diclofenac from spray-dried silk sericin microparticles were observed.	[42]
Silkworm cocoon	Spongy hydrogel prepared using silk sericin indicated random coil conformation.	[43]
Silkworm cocoon	Differently degummed silkworm sericin showed significant difference in tensile strength and fractography.	[44]
Silkworm cocoon	Sericin incorporated cell culture media accelerated the proliferation of murine hybridoma 2F3-O, human hepatoblastoma HepG2, human epithelial HeLa, human embryonic kidney 293 cells.	[45]
Silkworm cocoon	Silk sericin capped and *Rhodomyrtus tomentosa* leaf ethanolic extract reduced silver nanoparticles. Incorporated hydrogel demonstrated potential antimicrobial application.	[46]
Silkworm cocoon	Dissolution of sericin-alginate enteric-coated mucoadhesive particles fortified with diclofenac sodium protected the degradation in gastric acidic medium.	[47]

(Table 1) cont.....

Biopolymer	Application	Reference
Silkworm cocoon	Potential of sericin as cryo-protectant during lyophilization of human mesenchymal stromal cells and human osteoblast cells were-tested successfully.	[48]
Silkworm cocoon	Silver nanoparticles coated silk surgical sutures indicated biocompatibility with suitability in the management of surgical site infections considering their excellent bactericidal effects.	[49]

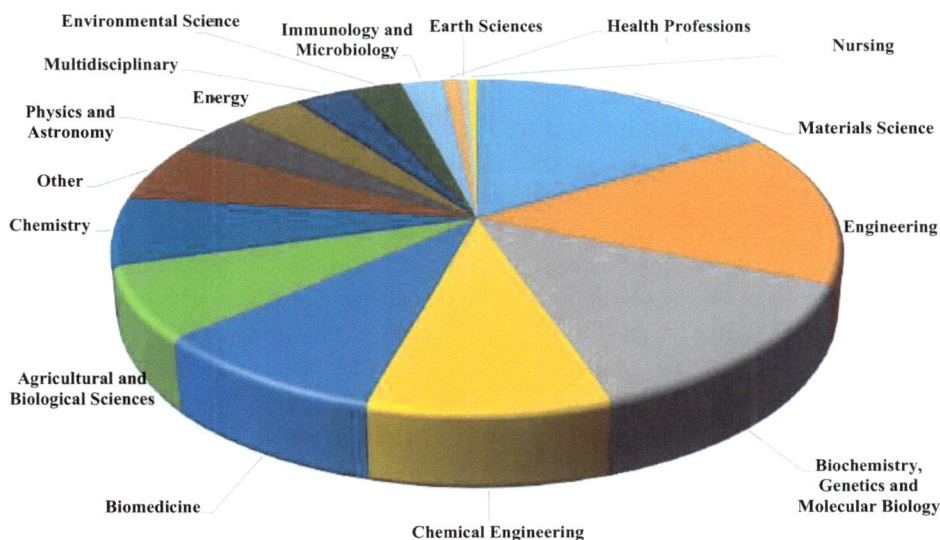

Fig. (3). Application of biopolymer silk sericin in the various fields available online on the Scopus database. The search includes Bullen terms such as "Silkworm cocoon*", "silk sericin*", "biopolymer*" (Assessed 23rd Aug 2021) (for interpretation of the results to color in this Fig. legend, the reader referred to either web version of this chapter or color print).

Biopolymers from Peptide and Polypeptides

Synthetic and natural polymers are widely applied in material science and biointerface engineering including tissue engineering, fabrication of drug delivery systems, and biosensors. Generally, natural biopolymers used in the development of pharmaceuticals include cellulose, collagen, hyaluronic acid, sericin, and fibroin. However, recently engineered bio-peptides have attracted significant attention among research scientists. Peptides and polypeptides are emerging natural biopolymers due to their unique physicochemical properties. The development of peptide and polypeptide-based biomaterials is preoccupied by the combination of engineered proteins and macromolecular assembly.

Peptides are specific proteins fragments that influence the body functions *via* a metabolic process of living organisms and consequently affect human health.

Moreover, peptides are organic materials formed by covalent bonding (amide or peptide bond) of amino acids, while proteins are polypeptides with a higher molecular weight [50]. These bioactive peptides and proteins play an important role in metabolic activities to demonstrate hormone or drug-like actions. Further, considering the metabolic process, peptides and proteins are classified based on the mode of actions as antimicrobials, antithrombotic, antihypertensive, opioid, immunomodulatory, mineral binding, and antioxidants. Furthermore, proteins obtained from amino acids *via* synthesis within an organism are classified as endogenous, while if obtained through the diet or from external sources to an organism including foods are known as exogenous [51] (Fig. **4**). The pharmacological efficacy and pharmaceutical applications of peptides and polypeptides depend on their structure, amino acid composition, type of *N*- and *C*-terminal amino acid, with the chain length, charge of amino acid forming peptides, *etc*. Several bioactive peptides and their engineered biomaterials were reported including poly-amino acids, elastin-like polypeptides, silk-like proteins, coiled-coil domains, tropoelastin-based peptides, leucine zipper-based peptides, peptide amphiphiles, beta-sheet forming ionic oligopeptide, beta-hairpin peptides, *etc*. Numerous studies reported that bioactive peptides and polypeptides are potentially used as antioxidants [52], antimicrobials [53], immunomodulators [54], cytomodulators [55], nutraceuticals, and as anti-inflammatory [52], antihypertensive [53], anti-genotoxic activity, anti-obesity [56], anti-cancer [54], and mineral binding agents as presented in Table **2**.

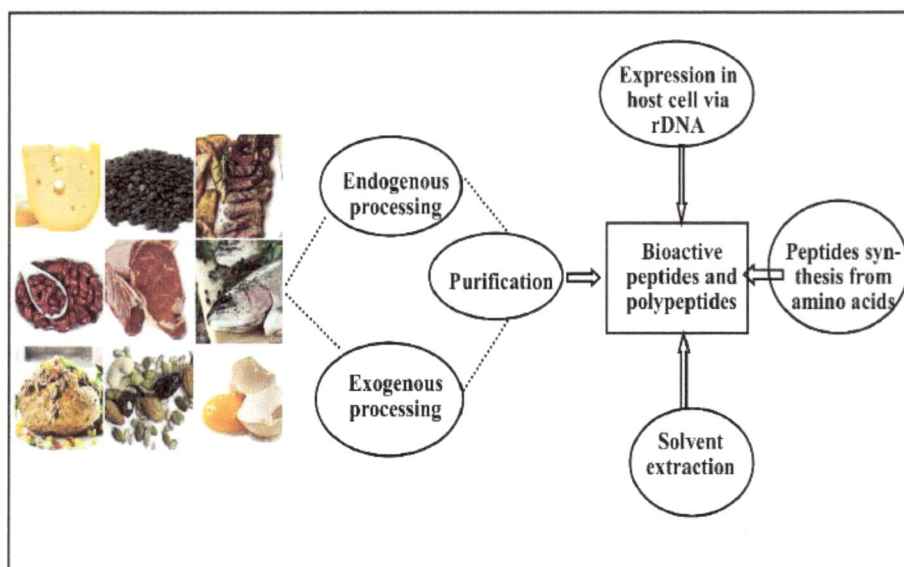

Fig. (4). Various ways to obtain bioactive peptides and polypeptides.

Table 2. Peptide and polypeptides as biopolymer functionalized novel drug delivery systems for the treatment of life threatening diseases.

Biopolymer (Peptides/Aptamer) Based Formulations	Application	Reference
Biological hydrogel consisting of ionic self-complementary peptides with 16 amino acids.	Controlled release scaffold for lipophilic drugs	[57, 58]
Short self-assembling peptides scaffolds and related modified variants coupled with short biological active motifs for the design of synthetic biological materials for 3D culture, reparative, regenerative medicine, and tissue engineering.	Enhancement in fibroblast migration *via* nanofibers	[59]
Hydrogel produced from self-assembling peptides and their peptide derivatives as extracellular matrices for cell culture substrates and scaffolds for regenerative medicine.	Endothelial cell proliferation	[60]
Hydrophobic polyphenol incorporated self-assembling peptide hydrogel.	Injectable hydrogel	[61]
Synthesis of a conjugated low molecular weight polylactide and related peptide for self-assembling structure and delivery of doxorubicin.	Drug release kinetics study *via* peptide assemble nanoparticles.	[62]
Nanoparticles encapsulated doxorubicin for targeted delivery to αvβ3-expressing tumor vasculature.	Nanoparticle mediated drug delivery to tumor.	[63]
Ribonucleic acid aptamer functionalized poly(lactic-c--glycolic acid)-lecithin-curcumin-polyethylene glycol nanoparticles for targeting human colorectal adenocarcinoma cells.	Aptamer fortified nanoparticles to improve target delivery of novel chemotherapeutic agents to colorectal cancer cells.	[64]
Anti-nucleolin aptamer (AS1411) for site-specific targeting against tumor cells.	AS1411 aptamer –PLGA-lecithin-PEG nanoparticles as a potential candidate for targeted drug delivery.	[65]
Aptamer functionalized cisplastin-albumin nanoparticles indicated improvement in the drugs' therapeutic efficacy and tolerability.	Atamer functionalized nanoparticles enhanced antitumor effects and tolerability, compared with free cisplastin.	[66]
Aptamer (AS1411) functionalized albumin-based doxorubicin nanoparticles for effective targeting drug delivery and targeted cancer therapy.	Aptamer-modified nanoparticles exhibited favorable effects on the prevention of tumor progression and metastasis.	[67]
Aptamer functionalized curcumin-loaded human serum albumin nanoparticles for targeting breast cancer cells.	Targeted nanoparticles with potential candidature in the treatment of cancer cells (HER2)	[68]

Lignin an Aromatic Biopolymers

Lignin is the second most abundant biomaterial on the earth, generated as a waste product from the paper and ethanol industry. Conversely, the paper industry valorizes lignin in huge amounts to recover it as fuel, however due to a large amount of lignin generation, considered as a potential source of biopolymer. The biomass reserve is estimated at approximately $1.85 - 2.4 \times 10^4$ tons on the earth among that 20% is lignin [69]. The worldwide production of lignin is approximately 100 million tons/year with a value of $732.7 million in 2015 that is expected to reach $ 913.1 million by 2025 with a compound annual growth rate of 2.2% [70]. Whereas, currently 50-70 million tons of lignin alone are produced by the pulp and paper industry globally and estimated to increase by 225 million tons per year until 2030 [71]. A Swiss botanist Candolle A.P. used the term lignin for the first time in 1813, which means "wood", a major component of lignocellulose biomass [72]. Lignin is a valuable source of biomaterial and energy, obtained from wood as cellulose and hemicellulose (Table **3**). Moreover, lignin can be used in the production of carbon fibers, which has shown excellent replacement of automotive steel, due to its lightweight with good tensile strength [73]. A recent study reported that lignin-based carbon fiber with a value of 1.1 billion $ per 100 kilotons will reach $ 6.19 billion per tons by 2022 [70].

Table 3. Source of important natural fibers [76].

Fibers	Origin	Species	Fibers	Origin	Species
Abaca	Leaf	*Musa textilis*	Istle	Leaf	*Samuela carnerosana*
Agave	Leaf	*Agave Americana*	Jute	Stem	*Corchorus capsularis*
Alfa	Grass	*Stippa tenacissima*	Kapok	Fruit	*Ceiba pentranda*
Bamboo	Grass	Several species (> 1250)	Kenaf	Stem	*Hibiscus cannabinus*
Banana	Leaf	*Musa indica*	Kudzu	Stem	*Pueraria thunbergana*
Broom root	Root	*Muhlembergia macroura*	Mauritius hemp	Leaf	*Furcraea gigantea*
Cadillo	Steam	*Urena lobate*	Nettle	Stem	*Urtica dioica*
Cantala	Leaf	*Agave Cantala*	Palm oil	Fruit	*Elaeis guineensis*
Caroa	Leaf	*Neoglaziovia variegate*	Phormium	Leaf	*Phormium tenas*
China jute	Stem	*Abutilon theophrasti*	Piassava	Leaf	*Attalea funifera*
Coir	Fruit	*Cocos nucifera*	Pineapple	Leaf	*Ananas comosus*
Cotton	Seed	*Gossypium spp.*	Ramie	Stem	*Boehmeria nivea*
Curaua	Leaf	*Ananas erectifolius*	Rosella	Stem	*Hibiscus sabdariffa*
Date palm	Leaf	*Phoenix dactylifera*	Sansevieria	Leaf	*Sansevieria*
Flax	Stem	*Linum usitatissium*	Sisal	Leaf	*Agave sisalana*

(Table 3) cont.....

Fibers	Origin	Species	Fibers	Origin	Species
Hemp	Stem	*Cannabis sativa*	Sponge gourd	Fruit	*Luffa cylindrical*
Henequen	Leaf	*Agave fourcroydes*	Sun hemp	Stem	*Crorolaria Juncea*
Isora	Stem	*Helicteres isora*	Wood	Stem	Several species

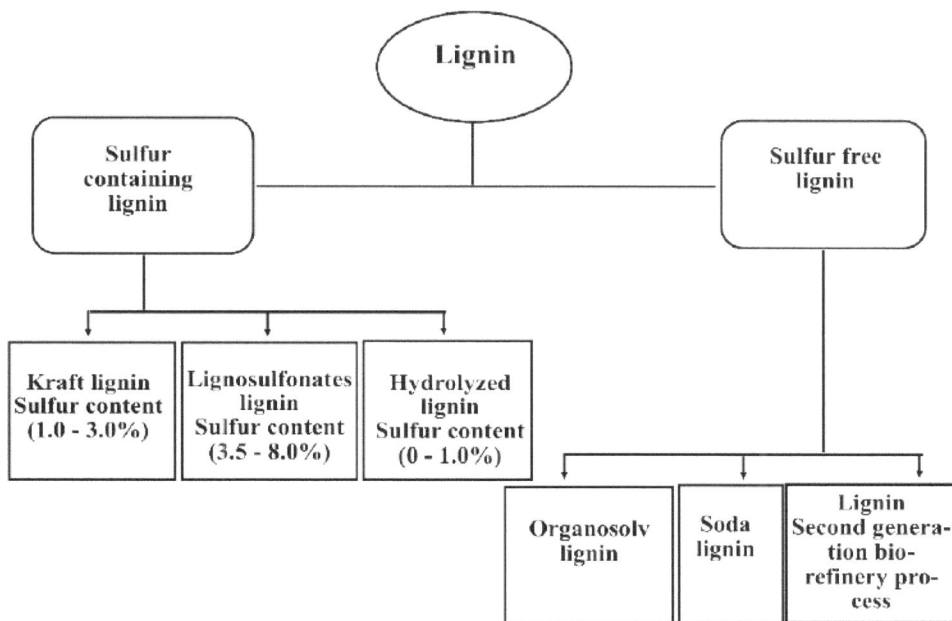

Fig. (5). Extraction of lignin with or without sulfur content in percentage [70].

The structure of lignin differs based on the extraction process and the presence of functional groups including hydroxyl, methoxyl, carbonyl, and carboxyl (Fig. **5**). Moreover, lignin is the most abundant form of natural aromatic biopolymer with a branched 3-D phenolic structure including phenylpropane monomers (*p*-coumaril, coniferyl, and sinapyl) also known as monolignols [74]. Hardwood lignin contains guaiacylsyringyl, whereas softwood has relatively low sinapyl units. In addition, lignin is categorized as lingo-sulphonate, kraft lignins, and organosolv lignins with the availability of 88%, 9%, and 2%, respectively [75]. Furthermore, to develop a formulation using novel lignin-based biopolymers, it is essential to perform controlled degradation to get uniform structural lignin. The demand and supply of paper with future dependency on papers are slowly disappearing due to a sharp surge in online data communication and e-mail services. This reduction in dependency on paper is an alarming situation for an investor, however, it could create an alternative opportunity for the paper industry to develop a lignin-derived product that could fit with the scale-up and other use. Fig. (**6**) demonstrates how the researcher is moving forward by addressing the drawbacks of lignin processing and utilizing it for various applications.

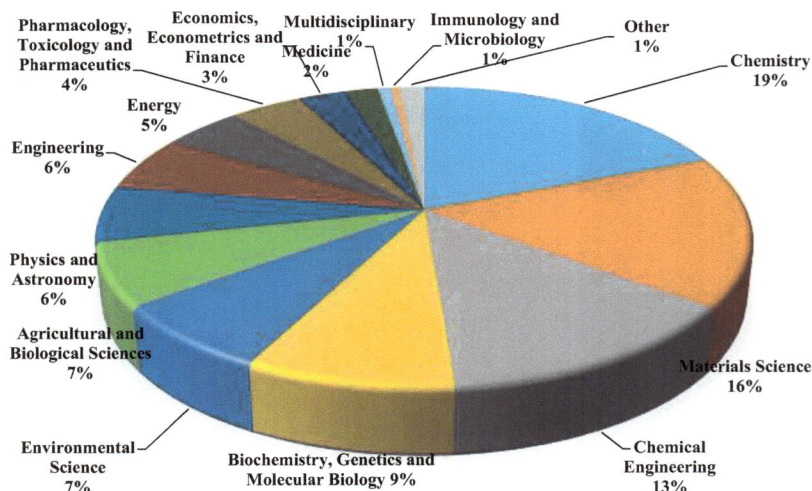

Fig. (6). Application of lignin as a biopolymer in the various fields available online on the Scopus database. The search includes Bullen terms such as "lignin*", "aromatic lignin*", "plant lignin*" (Assessed on 25th Aug 2021) (for interpretation of the results to color in this Fig. legend, the reader referred to either web version of this chapter or color print).

Recently, lignin-based research and new product development have gained significant momentum due to the progressive development of the biodegradable polymer concept, as aging pulp and paper mills that need to diversify products portfolio to maintain their vitality. Furthermore, effective "upstream" and "downstream" lignin valorization techniques also generated a demand for lignin as a binder and adhesive with potential application as bioplastics, concrete admixtures, and in biomedical practices. Several studies reported the possible usage of lignin as a biodegradable polymer, however the application of lignin-based formulation has yet not been commercialized. Table **4** highlights the potential pharmaceutical application of lignin and its derived materials with functional properties.

Table 4. Potential pharmaceutical application of lignin-derived formulations.

Application of Biopolymer Lignin	Reference
Methyl acrylate grafted lignin conjugated with polyvinyl alcohol encapsulated betulinic nano-formulation could be potential and cost-effective in the treatment of atherosclerosis.	[77]
Eco-friendly synthesis of lignin medicated silver nanoparticles indicated an effective sensor property against catalytic material within environmental remediation and alternative biomedical application.	[78]

(Table 4) cont.....

Application of Biopolymer Lignin	Reference
Enzymatically fabricated lignin colloidal particles for multifarious application in medicine, foods, and cosmetics.	[79 - 81]
The effects of lignin residues in the cellulose were examined for the properties and cytocompatibility of resultant hydrogel.	[82]
Composites fortified with green synthesized silver nanoparticles using lignin macromolecules demonstrated antibacterial efficacy against a wide range of microorganisms.	[83]
Chitin nanofibers complexed with nano-lignin and vitamin E incorporated biodegradable film demonstrated efficacy as beauty film.	[84]
Lignin and cellulose nanofibers fabricated microparticle nanocomposites indicated a promising alternative to fossil resource-derived microbeads for cosmeceutical applications.	[85]
Lignin based doxorubicin-loaded oil-in-water microemulsions with enhanced bioavailability and cytocompatibility indicated the potential to overcome the pharmacokinetic limitations.	[86]
Lignin with polyamidoamine dendritic polymeric nanofibers mats indicated enhancement of mechanical and thermal characteristics with potential application in drug delivery and filtration.	[87]
Polycaprolactone coated lignin-chitin fibrous gel as a candidate for controlled drug release based wound dressing.	[88]
Lignin capped silver nanoparticles fortified within polyurethane foam showed antibacterial and wound healing efficacy.	[89]
Lignin alginate-based microparticle beads presented a larger particle size with higher encapsulation and regulated release of atrazine for a prolonged duration.	[90]
Lignin-cellulose blend evaluated as a pharmaceutical excipient for manufacturing of tablets *via* direct compression.	[91]
Alkali-extracted lignin demonstrated improvement in stiffness of collagen gels without cytotoxicity and immunogenicity.	[92]
Lignin-nanoparticles network containing triterpenoids indicated lethal effects on tested cancer cells with thermal stability.	[93]
Lignin and cellulose were assessed as functional polymers in tableting technology *via* dry granulation techniques following quality by design.	[94]
Lignin-treated nanoparticles coated with lysine, glutamic acid, and transglutaminase as soft adhesive materials for dressing were fabricated for improvement in mechanical property.	[95]
Lignin incorporated with bacterial cellulose and dehydrogenative polymer hydrogel composite showed bactericidal effects against pathogenic bacteria.	[96]
Lignin containing self-nano emulsifying drug delivery of trans-resveratrol demonstrated enhanced oral absorption and stability.	[97]

(Table 4) cont.....

Application of Biopolymer Lignin	Reference
Lignin-based nanoparticles were fabricated using alkaline lignin with ethylene glycol followed by acidic treatments. The nanoparticles demonstrated good antioxidant and antimicrobial activity against potentially pathogenic bacteria.	[98]
Surface modified lignin nano-fibrous gel with arginine *via* electrostatic interaction accelerated the wound closure, re-epithelialization, collagen deposition, and angiogenesis.	[99]
Lignin and lignin modified nanoparticles showed promising platforms for different biomedical application.	[100]
Novel biodegradable nanoparticles from lignin were fabricated *via* sulfonated and acidic precipitation that could find application in drug delivery vehicles, stabilizers of cosmetic and pharmaceutical formulations.	[101]
Nano-scaled lignin *via* supercritical anti-solvent indicated improvement in the antioxidant potential.	[102]
Lignin-starch biodegradable film indicated improvement in mechanical property with the increments in lignin concentration and immediate release of drug.	[103]
Lignin-cellulose based hydrogel showed lignin concentration-dependent release of incorporated polyphenol extracted from grape seeds.	[104]
Imidacloprid lignin-polyethylene glycol matrices coated with ethylcellulose enhance the efficacy of the bioactive material and minimize the risk of environmental pollution.	[105]
Fabrication of polymerized alkali lignin, polyethylene glycol, and hexamethylene diisocyanate film with high solid content showed excellent mechanical strength with potential application in the coating of fertilizers.	[106]
Size controlled kraft lignin mediated nanoparticles with polyvinyl alcohol presented significant toughness and strength due to improved crosslinking bonds.	[107]
Biogenic gold nanoparticles *via* use of lignin and pulsed laser techniques showed excellent biosensor efficacy towards the detection of lead ions.	[108]
Lignin capped silver nanoparticles demonstrated significant antibacterial efficacy towards multi-drug resistant bacteria.	[109]
Kraft lignin medicated magnetic nanoparticles *via* pH-driven precipitation techniques presented potential application in the separation and removal of organic dye and heavy metals.	[110]
3D-printed lignin-based bio-composite fabricated using polyethylene glycol-2000 was able to improve tensile strength with good extrudability and printability, compared with struktol-TR451 included composite.	[111]
Demethylated lignin as wood adhesive showed good wood joint with dry shear strength at a low cost.	[112]
Methylated lignin grafted with dodecyl glycidyl ether composite showed high hydrophobicity and potential application in coating with packaging.	[113]
Phenolated lignin-based epoxy thermoset resins demonstrated good thermomechanical performance.	[114]

Co-processed Bio-Based Polymers

Solid oral tablets are widely acceptable and represent the most effective way to administer drugs to the patient. The formulation of cost-effective tablets includes the addition of various inactive pharmaceutical adjuvants together with active pharmaceutical ingredients. However, some specific solid oral tablets demanded to be designed with some particular criteria including dissolution, disintegration, and hardness with a regulated release. To meet such specific conditions, the formulation of solid oral tablets either by wet granulation or by direct compressions requires the incorporation of multifunctional polymers within the dosage form. However, the mixing of several functional polymers affects the cost of production. Therefore, a drug formulation scientist needs a multifunctional excipient that can effectively fulfill the requirements of two or three individual polymers. Recently several advanced techniques implemented to generate multifunctional polymers including copolymerization *via* grafting of polymer and processing of two excipients by co-precipitation, co-crystallization, solvent evaporation, hot-melt extrusion, agglomeration, dehydration, spray drying, *etc*.

Co-processing is particle engineering based on the concept of excipients interaction between particles at the micron level (Fig. **7**). The molecular levels comprise the arrangement of individual molecules in the crystal lattice including polymorphism, pseudo-polymorphism, and the amorphous state. While particle and bulk level comprise particle properties including shape, size, surface area, porosity, and density and flowability, compressibility, and dilution potential. Moreover, co-processed excipients are the combination of two or more compendial or non-compendial excipients using an appropriate method that can produce a synergistic modification in the function of polymer. The co-processed excipients were developed primarily to address the issues of flowability, compressibility, and disintegration potential with filler-binder adjuvant efficacy. Furthermore, the selection of excipients is an important aspect for co-processing considering compatibility with each other and retention or improvement of the desired property. The co-processed excipients are generally considered safe if the processed native excipients are certified by regulatory agencies [115]. Interestingly such a modification in inherent functional polymers represents an alternative to the production of novel inactive entities that might not require regulatory approval from the Food and Drug Administration. Table **5** and **6** represent the data of commercially available and patented co-processed excipients with their trade name available in the market. Moreover, processed polymers comply with the definition of excipients given by the International Pharmaceutical Excipients Council. The International Pharmaceutical Excipients Council represents the United States of America, Europe, and Japan that harmonize and regulate the requirements for purity and functionality testing of excipients. In addition, the processing of excipients no longer maintains the concept of inactive sustenance due to their modified biopharmaceutical aspects and technological impact. However, in some instances, improper processing of excipients might lead to the formation of salts. Table **7** highlights the bio-based co-processed potential pharmaceutical polymers with functional properties.

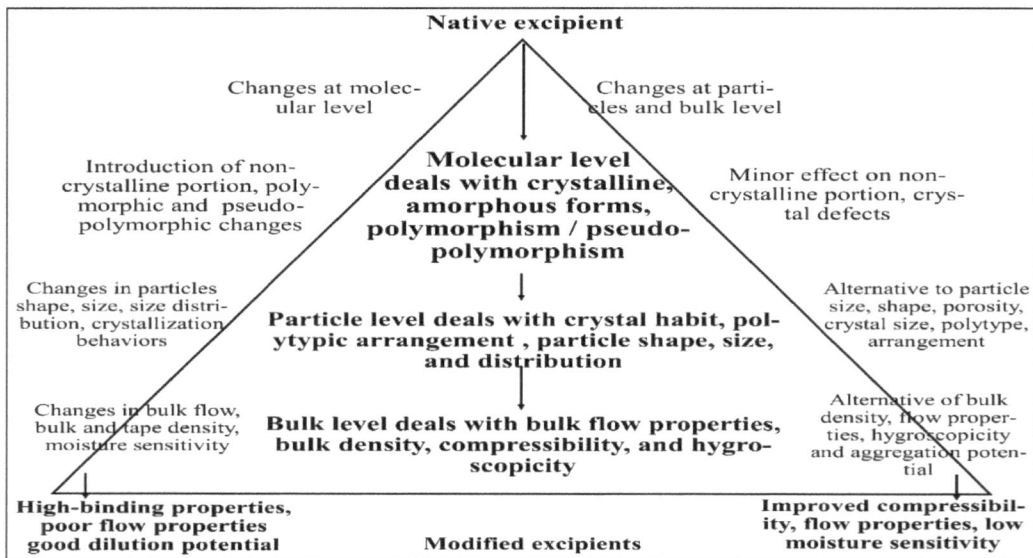

Fig. (7). Changes at various levels of solid-state pharmaceutical adjuvants during co-processing.

Table 5. Commercial co-processed excipients are available in the market as pharmaceutical adjuvants with their trade name.

Co-processed Excipients	Manufacturer	Trade Name
Lactose monohydrate 93.4, Kollidon 30-3.2%, Kollidon Cl-3.4	BASF SE Ludwigshafen, Germany	Ludipress®
Mannitol, Kollidon CL-SF-5, Kollidon SR30D-5	BASF SE Ludwigshafen, Germany	Ludiflash®
Lactose with cellulose	Meggle GmbH & Co. KG, Germany	Cellactose
Microcrystalline cellulose and silicon dioxide	Penwest Pharmaceuticals Company	Prosolv
Microcrystalline cellulose and guar gum	FMC Corporation	Avicel CE-15
Calcium carbonate sorbitol, microcrystalline cellulose, and lactose	Merck Meggle	ForMaxx
Lactose monohydrate with native corn starch	Merck Meggle	Starlac
B-lactose and lactitol	Dmvveghel	Pharmatose dcl 40

(Table 5) cont.....

Co-processed Excipients	Manufacturer	Trade Name
Xylitol-98, sodium carboxymethyl cellulose	Danisco	Xylitab
Mannitol, sorbitol, crospovidone, *silica*, aspartame, magnesium stearate	SPI Pharma	Pharmabust™ 500
Microcrystalline cellulose, hydroxypropyl methylcellulose, crospovidone	J.T. Baker	PanExcea™MC200G
Anhydrous lactose, glyceryl monostearate	Kerry Bio-functional Ingredient	LubriTose AN
Kollidon VA 64 and plasdone S630	Ashland	Copovidone
Ethoxylated hydrogenated castor oil	Seppic	Sepitrap 4000
Polysorbate 80	Seppic	Sepitrap 80
Carbohydrate, disintegrant, dicalcium phosphate	Fuji Chemicals	F-Melt
Microcrystalline cellulose and calcium carbonate	FMC Biopolymer	Vitacel VE-650
Tartaric acid	Pellet Pharmaceutical	Tap 400
Amorphous magnesium aluminometa*silica*te	Fuji Chemicals	Neusilin
Dicalcium phosphate anhydrous	Fuji Chemicals	Fujicalin
Directly compressible lactito	Cultor Food Science	Finlac DC
Starch and fructose	SPI Polyols	Advantage FS-95
Granulated mannitol	Roguette	Pearlitol SD
Sucrose and dextrin	Penwest Pharmaceutical	Dipac
Microcrystalline cellulose and lactose	Meggle GmbH Germany	Microcelac
Microcrystalline cellulose, colloidal silicon dioxide, crospovidone	RanQ Pharmaceuticals	Ran Explo-C
Microcrystalline cellulose, colloidal silicon dioxide, sodium starch glycollate	RanQ Pharmaceuticals	Ran Explo-S

Table 6. Patented co-processed excipient.

Co-processed Excipient	Patent Number
Microcrystalline cellulose and calcium carbonate	US 744987
Microcrystalline cellulose and galactomannan gum	US 5686107
Galactomannan with glucomannan	WO 95/17831
Mannitol with sorbitol	WO 2003/051338
Microcrystalline cellulose with disintegrant	WO 2010/132431A1
Vinyl lactam derived polymer with de-agglomerated materials	WO 2014/165246A1

Table 7. Recent highlights on the processing of biodegradable natural polymers with pharmaceutical applications.

Biopolymer	Application	Reference
Maltodextrin	Ionic liquid's active pharmaceutical ingredient *via* spray drying enabled conversion into a single and two-phase solid form.	[116]
Suweg starch	Effervescent tablets of Ibuprofen-PEG 6000 in solid dispersed form with co-processed suweg starch obtained from *Amorphophallus paeoniifolius*, lactose suggested compatability and no interaction among drug and polymers with the suitability of processed excipients in the development of tablets that meet the standard guidelines.	[117]
Buchanania lanzan	*Buchanania lanzan* seed polysaccharides co-spray dried with mannitol, lactose, and silicon dioxide indicated reduced cohesiveness on packability index and compressibility with initial fragmentation followed by plastic deformation.	[118]
Manilkara zapota	*Manilkara zapota* seed polysaccharides co-processed with hypromellose, sodium carboxymethyl cellulose, and polyvinylpyrrolidone indicated good work of adhesion with biocompatibility.	[119]
Chitosan	Co-processed chitosan chlorhydrate with mannitol fabricated following the design of experiment showed good optimization competence in the development of directly compressible excipients.	[120]
Cellulose	Effect of atomization pressure and polymer ratio was evaluated using the design of experiments for co-processed multifunctional excipients.	[121]
Hypromellose	Modified release system for active constituents of *Physalis peruvia*na was processed with hypromellose resulting in regulated release and hypoglycemic effects.	[122]
Chitin and starch	Magnesium *silica*te co-processed on chitin, microcrystalline cellulose or starch demonstrated improvement in flowability, compactibility, stability with multifunctional applications in tableting technology.	[123]
Gelatin and cellulose	Co-processed gelatin or microcrystalline cellulose with tapioca starch and colloidal silicon dioxide, suggested that the choice of binder used during the processing of excipients plays a crucial role in tableting technology.	[124]
Triglycerides	Supersaturated self-nanoemulsifying pre-concentrate of fenofibrate using novel co-processed excipients developed from dimethyl acetamide, triglycerides, and kolliphor significantly enhanced the dissolution of lipophilic drug.	[125]
Starch	Co-processed mannitol with starch or povidone indicated improvement in compactibility and compressibility.	[126]
Chitosan	Co-processed crab chitosan with three grades of prosopis gum resulted in a concentration dependent viscosity and adhesive strength.	[127]
Lactose	The combination of fibrous microcrystalline with co-processed lactose indicated suitability in development of multiple-unit pellets.	[128]

(Table 7) cont.....

Biopolymer	Application	Reference
Phoenix dactylifera	Directly compressible excipients from *Phoenix dactylifera* and microcrystalline cellulose resulted in the improvement of mechanical and disintegration properties.	[129]
Hypromellose	Co-processed ibuprofen with various grade of hypromellose resulted in the improvement of solubility and intestinal permeability.	[130]
Hypromellose	Co-processed hypromellose with lactose and sodium chloride exhibited improvement in dispersibility and dissolution rate.	[131]
Cellulose and chitosan	Co-processed microcrystalline cellulose, sorbitol, chitosan, and Eudragit resulted in the enhancement of solubility and dissolution of BCS II drugs.	[132]
Cellulose	Co-processed sodium carboxymethyl cellulose and silicon dioxide with plant extract of *Hamamelis virginiana* demonstrated improved flow-ability and suitability in the development of modified-release hydrophilic matrix.	[133]
Lactose	Spray dried lactose presented improved physicomechanical and cushioning effects.	[134]
Starch	Co-processed maize starch and carbopol 974P with metoprolol tartrate nasal spray resulted in improved bioadhesion and bioavailability.	[135]
Cellulose	Microcrystalline cellulose co-processed with colloidal silicon dioxide, mannitol, fructose, and cross povidone enhanced dry binding potential with reduction in disintegration time.	[136]

Graft Co-Polymerizations

Polymers play an essential role in the development of drug delivery that could fit with the needs of the patient in the modern world. Traditionally researchers in medical science view polymers as an integral component of not only therapeutics but also utilized them in the fabrication of several devices including inhalers, catheters, bio-prostheses, transdermal patches, *etc*. Moreover, recently polymeric materials have been investigated for a safer way to deliver genes. In addition, polymers have also been used for biosensors, testing devices, bioregulation, and insulation. Furthermore, polymeric materials incorporated in textile products also demonstrate excellent aesthetic properties. In short, polymers play a role in every aspect of our lives. Although polymers are legion, sometimes they could not fulfill the demand, depending on their properties. Therefore, advancements in polymers are reckoned necessary to widen the scope of their applications either by the development of new molecules or *via* modification in the property of existing polymers. The development of new polymeric molecules is a tedious process that involves mandatory regulatory approvals. However, modification in the existing polymeric materials is an economic process that opens new possibilities for improvement in surface and bulk properties.

The process of linking similar blocks of monomers forms a polymeric material, however, processing of the monomers of different kinds produces a product known as a copolymer, and the process of uniting them is called copolymerization. Moreover, depending on the different monomer chain lengths and other properties, the copolymer products are diversified in their property profiles without disturbing biodegradability. Generally, copolymerization with an alternative sequence of monomers forms alternating copolymers, while random copolymers do not follow any definite sequence in alignment patterns of native monomers. Several techniques are available for polymer modification including crosslinking, blending, and grafting. Graft copolymerization produces branched macromolecules that significantly differ in branches types, compared to native monomers backbone [137]. Whereas, polymerization of monomers *via* irradiation or sulfur vulcanization leads to the formation of a network structure, defined as crosslinking [138]. Furthermore, a homogenous mixture of two or more arbitrary different poly-dispersed polymers is defined as blending. The copolymer grafting is carried out *via* various chemical reactions including redox reaction, radical formation, enzymatic reaction, gelation, and using phenolic, amino, and epoxy resins (Fig. **8**).

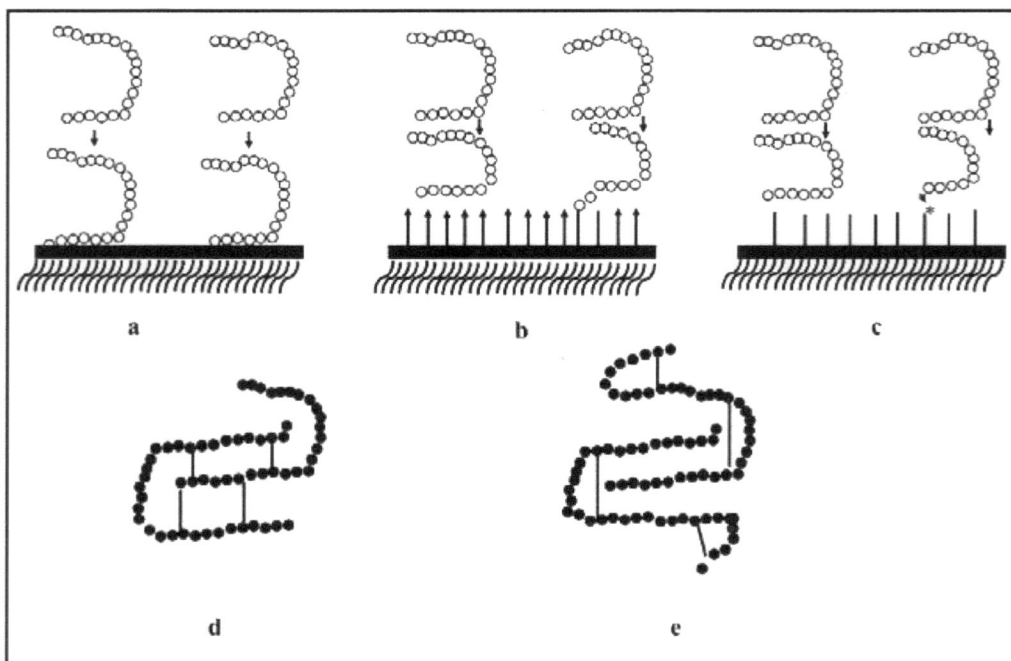

Fig. (8). Schematic presentation of physio-sorption (**a**), initiation of grafting (**b**), grafting formation (**c**), intermolecular crosslinking (**d**), and intramolecular crosslinking [140].

Common grafting of monomers or two different monomers is carried out either by redox reaction or by free radical formations. Redox reactions are initiated *via* Fe^{2+}/H_2O_2, Fe^{2+}/hydroperoxides, Fe^{2+}/Persulphate, persulphate with a reducing agent (sodium bisulphite, thiosulphate, Ag^+, *etc.*), oxidation, metal chelation, and by some indirect methods *via* generation of secondary free radicals. Later on, redox reaction occurs followed by propagation and termination. Whereas, free radical formation skips the chain termination stage in two ways, either by increasing the rate of initiation or by eliminating chain termination and transfer reactions [139]. Recent research highlights graft-copolymerization of bio-based polymers presented in Table **8**.

Table 8. Graft copolymerization of bio-based excipients with their pharmaceutical and application in food industry.

Biopolymer	Application	Reference
Lannea coromandelica	Biopolymer grafted polyacrylamide indicated higher swelling at neutral pH with controlled release of active pharmaceutical for suitability as colon targeted drug delivery.	[141]
Corn fiber	Polyacrylamide grafted corn fiber gum showed shear thinning behavior that followed Herschel Bulkley kinetic model with potential as a functional polymer to be used in food and pharmaceuticals.	[142]
Karaya gum	Gum karaya grafted with dimethylamino ethyl methacrylate and *N',N'*-methylene-bis-acrylamide gel revealed excellent dye adsorption efficacy.	[143]
Cellulose	Cellulose biopolymer functionalized by free graft copolymerization and used as adsorbent to eliminate toxic inorganic pollutant from waste water.	[144]
Cellulose	Graft-polymerization of 2-2-dimethyl-1-3-dioxolan-4-yl methyl acrylate on hydroxyethyl cellulose showed two fold of moisture absorption compared with native cellulose.	[145]
Starch	Improvement in mechanical property was observed for starch grafted with methacrylic acid film.	[146]
Starch	Hydrogel fabricated using graft starch with *N',N'*-methylene-b-s-acrylamide demonstrated excellent gelling capacity in water and salt.	[147]
Okra gum	Okra gum grafted with polyacrylamide showed excellent sustained release efficacy for incorporated drug.	[148]
Chitosan	Chitosan bead grafted with poly(methyl methacrylate) resulted in improvement in coating wall material property and swelling capacity.	[149]
Starch	Microwave assisted grafting of starch with Poly(N-acryloyl-L-phenylalanine) showed potential in delivering drugs in high alkaline medium.	[150]
Guar gum	Guar gum grafted with vinyl monomers demonstrated regulated release of entrapped molecules.	[151]

(Table 8) cont.....

Biopolymer	Application	Reference
Cellulose	Surface graft copolymerization of cellulose showed potential application in paper industry.	[152]
Slap	Graft copolymerized slap used as a reducing and capping agent in the synthesis of silver nanoparticles.	[153]
Chitosan	pH sensitive porous chitosan membrane was prepared *via* surface graft copolymerization in supercritical carbon dioxide.	[154]
Cellulose	Hemicellulose grafted and copolymerized with acrylic acid and acrylic amide based hydrogel demonstrated excellent water absorbency property.	[155]

Biodegradable Polymers From Marine Origin

Water covers almost three-fourth of the globe's surface with the availability of a wide range of marine organisms that supply or are used in the development of valuable biomaterials including lipids, proteins, and polysaccharides. Moreover, marine microorganisms such as bacteria, microalgae, and seaweed represent a large source of other valuable biodegradable polymers. The seaweed polysaccharides are considered the most abundant and widely available in the ocean, from tidal level to free-floating or anchored. So far, biopolymers commercially exploited from the marine source are alginate, carrageenan, and agar. In addition, chitin is the second most plentiful carbohydrate-based polymer obtained from the waste product of crab, squid pens, and crayfish exoskeleton, which also come from the ocean. Recent highlights on various applications of biodegradable polymers isolated from various marine sources are presented Table 9.

Table 9. Pharmaceutical and allied application of biodegradable adjuvants extracted from various marine sources.

Biopolymer	Application	Reference
Carrageenan	Enhancement of solubility and oral bioavailability of curcumin fortified with ι-carrageenan nanoemulsion based composite.	[173]
Carrageenan	Insulin entrapped with caboxymethylated ι-carrageenan grafted chitosan showed improved oral bioavailability with extended glycemic control in insulin therapy.	[174]
Carrageenan	Graft polyvinyl alcohol-hyaluronic acid-g-κ-carrageenan hydrogel incorporated with ampicillin sodium hydrogel prepared by freeze thawing demonstrated excellent biocompatibility, antibacterial, and wound healing efficacy.	[175]
Carrageenan	Caboxymethylated κ-carrageenan with polyvinyl alcohol nanofibers with enhanced blood coagulation and antibacterial efficacy with superior wound healing applications.	[176]

(Table 9) cont.....

Biopolymer	Application	Reference
Carrageenan	K-carrageenan and konjac glucomannan composite fortified with titanium dioxide showed food preservation and packaging application with excellent antibacterial activity.	[177]
Agar	Melon juice stability with the extension of self-life was observed with concentration dependent fortification of agar.	[178]
Agar	Starch-agar biodegradable packaging film indicated controlled degradation with potential efficacy in food preservation.	[179]
Agar	Starch-agar film plasticized with sorbitol resisted water vapor permeability, compared with starch film alone.	[180]
Alginate	Gold nanoparticles synthesized *via* chemical oxidative polymerization using sodium alginate and polyaniline boronic acid showed deleterious antioxidant and antimicrobial activity with cellular biocompatibility.	[181]
Alginate	Sodium alginate encapsulated with *Eucalyptus camaldulensis* polyphenols' ethanolic extract spheres matrix preserved the efficacy of bioactive compound within the extract with antimicrobial and antioxidant activity.	[182]
Alginate	Silver nanoparticles incorporated with antibacterial peel-off mask formulated using polyvinyl alcohol with sodium alginate indicated excellent antibacterial and biocompatibility against fibroblast L929 cell lines.	[183]
Chitosan	Chitosan extracted from shrimp shells-polyvinyl alcohol film fortified with phenolic rich natural anthocyanin isolated from flowers and fruits was evaluated for freshness of beverages.	[184]
Chitosan	Chitosan extracted from shrimp shells fortified with phenolic rich natural anthocyanin isolated from riceberry was used for monitoring the freshness shrimp.	[185]
Chitosan	Chitosan-polyvinyl alcohol film fortified with metallic nanomaterial demonstrated excellent mechanical strength, antibacterial, and antioxidant activity with the extension of meat shelf life.	[186]

Carrageenan is a generic name for biodegradable polysaccharides obtained by extraction from certain species of red seaweeds derived from the class of Rhodophyceae (*Gigartina, Chondrus crispus, Eucheuma,* and *Hypnea*) [156]. The word carrageenan is derived from the colloquial Irish name carrageen means "little rock". The first aqueous extraction of hydrophilic colloids was performed in 1810 in Ireland from red seaweed [156]. Carrageenans are categorized into six basic forms including ι-, κ-, λ-, μ-, υ-, and θ. Among them, κ-carrageenan is primarily extracted from tropical seaweeds *Kappaphycus alvarezii* (*Eucheuma cottonii*), while ι-carrageen and λ-carrageen are isolated from *Eucheuma spinosum* and *Gigartina* (*Chondrus* genera), respectively. In addition, the seaweeds are extracted using alkaline pH solution with high temperature to transmute the bio-precursor of μ- and υ- carrageenan into commercial ι- and κ- carrageenan. Furthermore, the gametophyte plants produce κ-/ι- hybrid carrageenan and sporophyte plants of these seaweeds produce λ-carrageenan [157].

Carrageenans are hydrophilic linear sulphated galactans consisting of alternating 3-linked β-D-galactopyranose and 4-linked α-D- galactopyranose, developing disaccharides repeating unit of carrageenan. The sulphated galactans are categorized conferring the existence of 3,6-anhydro-bridge on the 4-linke--galactose residue with position and number of sulphate group. The κ-, ι-, and λ- carrageenan are dimers with one, two, and three-sulphate ester groups, respectively. Though κ-, ι-, and λ- carrageenans commercially contain 22% (w/v), 32% (w/v), and 38% (w/v) of sulphate, they might differ depending on extraction batches and seaweed species [158]. Some carrageenans additionally contain substituents such as *O*-methyl group on position 6 of D unit similar to κ-carrageenan and pyruvate groups at position 4 and 6 of 3-linked-galactose units similar to λ- carrageenan [159]. Moreover, the presence of terminal xylose in small quantities has also been reported in carrageenan [160]. Carrageenans are widely used in the food industry that account for 70-80% of total production throughout the world with a total market of $ 300 million/year. In addition, carrageenan is also used in non-food products due to nontoxic nature of pharmaceuticals, cosmetics, printing, textile fabrications, and low-cost source of drugs. Furthermore, several studies reported that carrageenan demonstrates excellent pharmacological activities including antitumor, immunomodulatory [161], antihyperlipidemic [162], anticoagulant [163], inhibition of hepatitis A replication [164], and inhibition of broad range of human genital papillomaviruses [165] with potential usage as a standard inflammatory agent in experimental inflammations [166]. Statistical analysis of published research papers on "carrageenan" and its derivatives is available online on the Scopus database as presented in Fig. (**9**).

Marine organisms represent an enormous reservoir of bioactive and biopolymers with several applications. Agar as rich cell wall polysaccharides obtained from the inner matrix of red algal belongs to the family of *Gracilariaceae, Gelidiaceae, Petrocladiaceae,* and *Gelidiellaceae.* Agar within red algal cell wall provides confrontation to the pathogen, maintains cellular ionic equilibrium, and protects algae from ocean salinity, pH, and temperature. Depending on the structural conformation and side chain, substitution of algal galactans is referred to as agarose, agaroids, agarans, and agaropectin. Agar is a seaweed hydrocolloid or polycocolloid, with a long history of application in pharmaceuticals as gelling, thickening, cosmetics, and stabilizing food additives with extensive use in biotechnological industries. Furthermore, studies indicated that agar exhibits several pharmacological activities including anticoagulant, antiviral, antioxidant, anticancer, and immune-modulation.

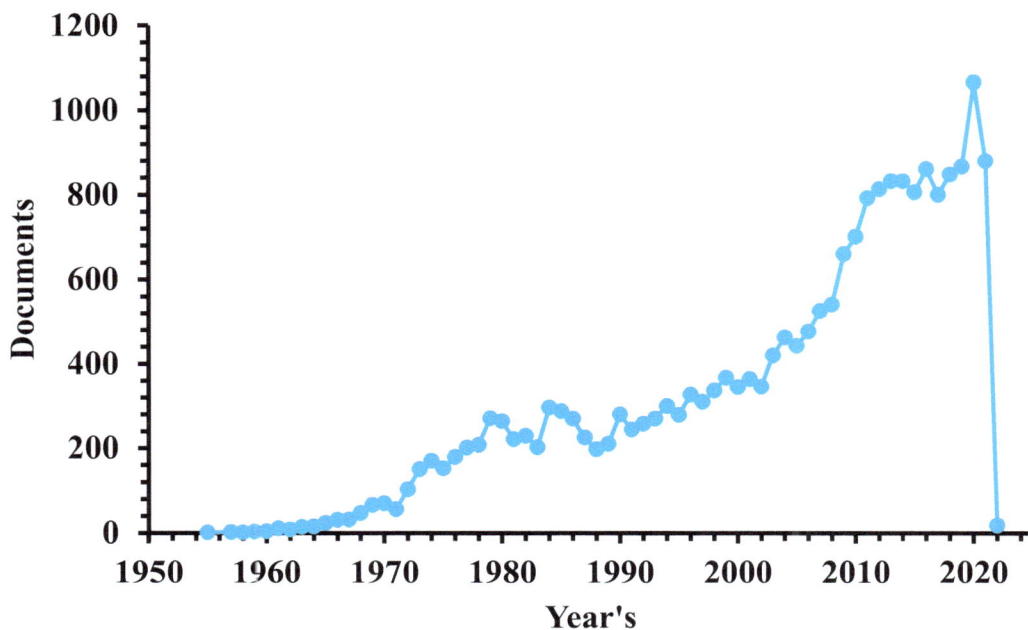

Fig. (9). Statistical analysis of published research papers on "carrageenan" and its derivatives available online on the Scopus database. The search includes Bullen terms such as "Carrageenan*", "biopolymer*" (Assessed on 20th Oct 2021) (for interpretation of the results to color in this Fig. legend, the reader referred to either web version of this chapter or color print).

The main source of alginate and its derivatives are species of brown algae (*Phaeophyceae*). The process for the extraction of "algin" from brown seaweeds was first reported and patented in 1881 by Stanford. The alginate is a salt of alginic acids produced from the marine algae which is a linear, anionic polysaccharide consisting of a variable amount of uronic acids, namely D-mannuronic acid (M) and L-guluronic acids (G), arranged in blocks of repeating M, G, and mixed MG residue. The biological function of alginate in brown algae is of a structure-forming agent. The intracellular alginate gel matrix gives mechanical strength and flexibility to the plant. However, the structural and functional differences vary among the alginates obtained from various species as well as from the marine region. The alginates are produced as a range of salts, however sodium salt is predominantly used for commercial purposes in food and pharmaceutical industries. Alginate is widely used for thickening, stabilizing, encapsulations, and gelling with good packaging and film-forming potential. Moreover, alginate in different salt forms is used to restructure the food, to prepare bakery cream, dessert jellies, and as fish feed. In addition, pharmacological studies have shown that alginic acid has an antianaphylaxis effect [167], immunomodulatory [168], and antioxidant activities [169], with anti-inflammatory [170], hypocholesterolemic, and antihypertensive efficacy. Moreover, free hydroxyl and carboxyl groups distributed along the backbone of alginate make it a suitable candidate for further structural and functional modifications to develop the product as demanded. Application of biopolymer alginate in various fields available online on the Scopus database is presented in Fig. (**10**).

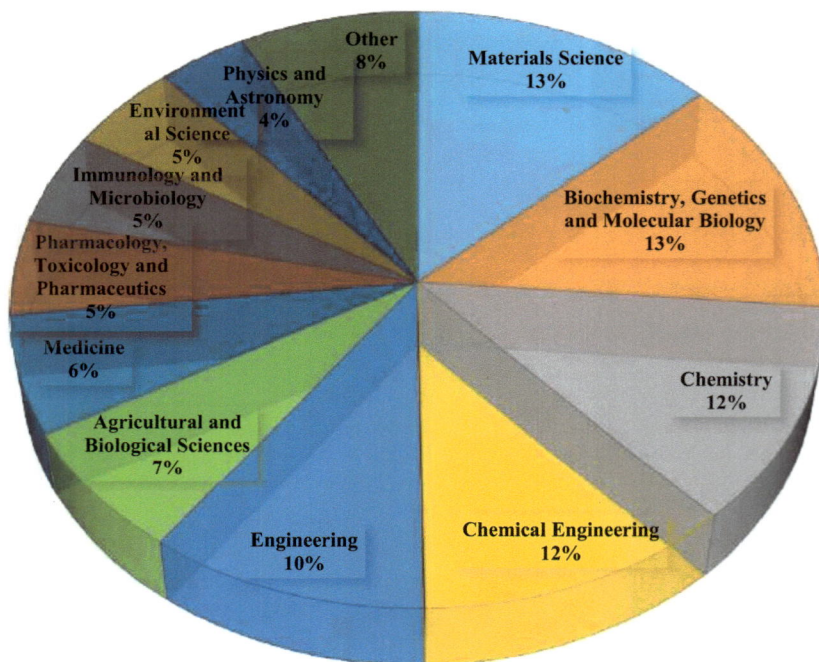

Fig. (10). Application of biopolymer alginate in various fields available online on the Scopus database. The search includes Bullen terms such as "alginate*", "alginic acid*", "biopolymer*" (Assessed on 20th Oct 2021) (For interpretation of the results to color in this Fig. legend, the reader referred to either web version of this chapter or color print).

Chitin is the most abundant natural amino mucopolysaccharide that can be easily available from the shell of crab or shrimp. Whereas, chitosan is a biodegradable and biocompatible derivative of natural chitin prepared *via* alkaline deacetylation of chitin with numerous applications in various pharmaceutical and food industries. The extraction of chitin involves an acid removal of calcium carbonate, using a hot reaction with HCl or HNO_3, followed by deproteinization by alkaline treatments. Structurally chitosan is comprised of β-1,4-linked 2-amino-2-do-y-β-D-glucose and *N*-acetyl-D-glucosamine units [171]. However, chitin structurally can be modified by removing the acetyl group bonded to amine radicals in the C_2 position on the glucan ring using hydrolysis in concentrated alkaline solution at an elevated temperature. Commercially, chitosan is available with > 85% deacetylated unit and molecular weight between 100 to 1000 kDa. Although chitosan is soluble in acidic solution, the chitosan-based products are used in numerous applications including wound healing, tissue engineering, gene delivery, protein binding, cell encapsulation, fabrication of implants, and contact lenses, bio-imaging, food additives, food packaging, and antibacterial textiles. Furthermore, several studies have reported the introduction of permanent positive charge in the polymeric chain to modify cationic polyelectrolyte, resulting from improvement in the aqueous solubility [172]. Thus, improvement in aqueous solubility of modified chitosan *via* methylation widens the applicability in pharmaceutical and other sectors. Statistical analysis of published research papers on "chitosan" available online on the Scopus database is presented in Fig. (**11**).

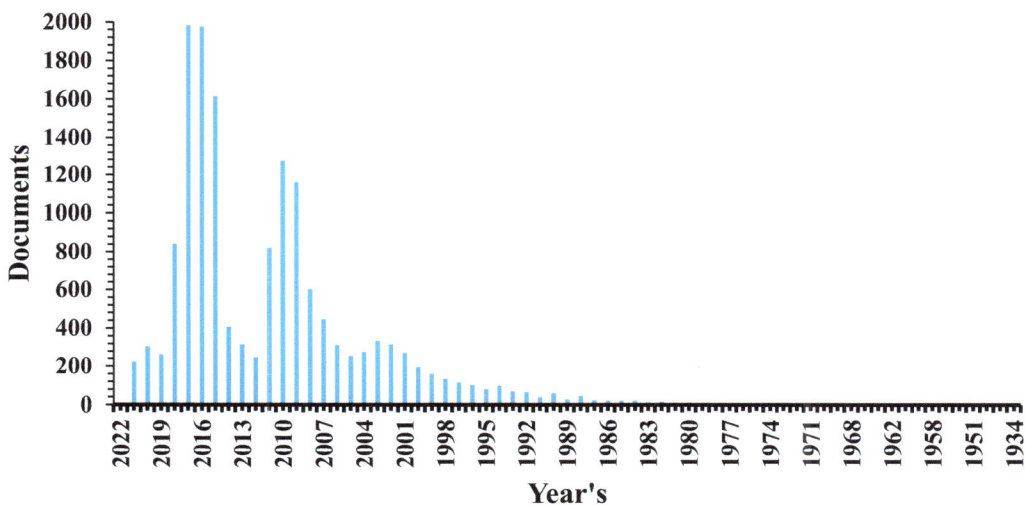

Fig. (11). Statistical analysis of published research papers on "chitosan" available online on the Scopus database. The search includes Bullen terms such as "chitin*", "chitosan*", "biopolymer*" (Assessed on 20th Oct 2021).

CONCLUSIONS AND FUTURE PROSPECTS

The use of biodegradable and processed polymers with increased functionality has attracted researchers since the mid of 19th century due to their biocompatibility and efficacy. The approach of novel and cost-effective targeted drug delivery is still in transition phase towards the comprehensive utilization of natural biodegradable polymers. Moreover, natural biodegradable polymers have been widely used by food industry too and in near future represent a valuable and low cost source of novel functional excipients. In addition, the increasing knowledge of structural modification within natural polymers developed thirst area in search of synergistic therapeutic biomaterials. Furthermore, structurally modified polymers are as similar to novel excipients that give certain hopes among investors to develop cost-effective multifunctional polymers with accepted regulatory norms. However, the different therapeutic applications of natural biodegradable and processed polymer are still in the experimental phase that might take time to reach public domain, and meanwhile, significant attention is also required on the application of natural polymers in several other fields that still hold a considerable promise for future utilization.

REFERENCES

[1]   Vroman I, Tighzert L. Biodegradable Polymers. Materials (Basel) 2009; 2(2): 307-44.
        [http://dx.doi.org/10.3390/ma2020307]

[2]   Chaabouni EGF, Brar SK. Biopolymers Synthesis and Application. In: Brar SDG, Soccol C, Eds. Biotransformation of Waste Biomass into High Value Biochemicals. New york: Springer 2014; pp. 415-43.
        [http://dx.doi.org/10.1007/978-1-4614-8005-1_17]

[3]   Singh B, Sharma N. Mechanistic implications of plastic degradation. Polym Degrad Stabil 2008; 93(3): 561-84.
        [http://dx.doi.org/10.1016/j.polymdegradstab.2007.11.008]

[4]   Chen F, Porter D, Vollrath F. Structure and physical properties of silkworm cocoons. J R Soc Interface 2012; 9(74): 2299-308.
        [http://dx.doi.org/10.1098/rsif.2011.0887] [PMID: 22552916]

[5]   Altman GH, Diaz F, Jakuba C, *et al.* Silk-based biomaterials. Biomaterials 2003; 24(3): 401-16.
        [http://dx.doi.org/10.1016/S0142-9612(02)00353-8] [PMID: 12423595]

[6]   Kirshboim S, Ishay JS. Silk produced by hornets: thermophotovoltaic properties—a review. Comp Biochem Physiol A Mol Integr Physiol 2000; 127(1): 1-20.
        [http://dx.doi.org/10.1016/S1095-6433(00)00237-3] [PMID: 10996813]

[7]   Akai H. Anti-bacterial function of natural silk materials. Int J Wild Silkmoth & Silk 1997; 3: 79-81.

[8]   Das S, Shanmugam N, Kumar A, Jose S. Review: Potential of biomimicry in the field of textile technology. Bioinspired, Biomimetic and Nanobiomaterials 2017; 6(4): 224-35.
        [http://dx.doi.org/10.1680/jbibn.16.00048]

[9]   Lee YW. Silk reeling and testing manual: Food Agri Org. 1999; 255-96.

[10]   Kundu SC, Dash BC, Dash R, Kaplan DL. Natural protective glue protein, sericin bioengineered by silkworms: Potential for biomedical and biotechnological applications. Prog Polym Sci 2008; 33(10): 998-1012.
        [http://dx.doi.org/10.1016/j.progpolymsci.2008.08.002]

[11]   Takasu Y, Yamada H, Tsubouchi K. Isolation of three main sericin components from the cocoon of the silkworm, *Bombyx mori.* Biosci Biotechnol Biochem 2002; 66(12): 2715-8.
        [http://dx.doi.org/10.1271/bbb.66.2715] [PMID: 12596874]

[12]   GAMO T. Electrophoretic analyses of the protein extracted with disulphide cleavage from cocoons of the silkworm. Bombyx mori L J Seric Sci Jpn 1973; 42(1): 17-23.

[13]   Gamo T, Inokuchi T, Laufer H. Polypeptides of fibroin and sericin secreted from the different sections of the silk gland in *Bombyx mori.* Insect Biochem 1977; 7(3): 285-95.
        [http://dx.doi.org/10.1016/0020-1790(77)90026-9]

[14]   Sprague KU. Bombyx mori silk proteins. Characterization of large polypeptides. Biochemistry 1975; 14(5): 925-31.
        [http://dx.doi.org/10.1021/bi00676a008] [PMID: 1125178]

[15]   Magoshi J. Biospinning (silk fiber formation, multiple spinning mechanisms). Polym Mat Encycl 1996; 1: 667-79.

[16]   Teli MD. Textile coloration industry in India. Color Technol 2008; 124(1): 1-13.
       [http://dx.doi.org/10.1111/j.1478-4408.2007.00114.x]

[17]   Lesile M, Stephen M, Robert S. Cotton and wool outlook. Econ Res Service. USDA, CWS 2003; 303(11): 1-15.

[18]   Kang YJ, Jo YY, Kweon H, *et al.* Comparison of the physical properties and *in vivo* bioactivities of flatwise-spun silk mats and cocoon-derived silk mats for guided bone regeneration. Macromol Res 2020; 28(2): 159-64.
       [http://dx.doi.org/10.1007/s13233-020-8026-z]

[19]   Da Costa TB, da Silva MGC, Vieira MGA. Crosslinked alginate/sericin particles for bioadsorption of ytterbium: Equilibrium, thermodynamic and regeneration studies. Int J Biol Macromol 2020; 165(Pt B): 1911-23.
       [http://dx.doi.org/10.1016/j.ijbiomac.2020.10.072] [PMID: 33091471]

[20]   Wu Z, Meng Z, Wu Q, *et al.* Biomimetic and osteogenic 3D silk fibroin composite scaffolds with nano MgO and mineralized hydroxyapatite for bone regeneration. J Tissue Eng 2020; 11
       [http://dx.doi.org/10.1177/2041731420967791] [PMID: 33294153]

[21]   Manjunath RN, Khatkar V. Investigation of structural morphology and mechanical behavior of muga silkworm (Antheraea assamensis) cocoons. J Nat Fibres 2020; pp. 1-13.

[22]   Miguel GA, Álvarez-López C. Extraction and antioxidant activity of sericin, a protein from silk. Braz J Food Technol 2020; 23: e2019058.
       [http://dx.doi.org/10.1590/1981-6723.05819]

[23]   Zhou CJ, Li Y, Yao SW, He JH. Silkworm-based silk fibers by electrospinning. Results Phys 2019; 15: 102646.
       [http://dx.doi.org/10.1016/j.rinp.2019.102646]

[24]   Akturk O, Gun Gok Z, Erdemli O, Yigitoglu M. One-pot facile synthesis of silk sericin-capped gold nanoparticles by UVC radiation: Investigation of stability, biocompatibility, and antibacterial activity. J Biomed Mater Res A 2019; 107(12): 2667-79.
       [http://dx.doi.org/10.1002/jbm.a.36771] [PMID: 31393664]

[25]   Tao G, Cai R, Wang Y, *et al.* Bioinspired design of AgNPs embedded silk sericin-based sponges for efficiently combating bacteria and promoting wound healing. Mater Des 2019; 180: 107940.
       [http://dx.doi.org/10.1016/j.matdes.2019.107940]

[26]   Dong Z, Guo K, Zhang X, *et al.* Identification of *Bombyx mori* sericin 4 protein as a new biological adhesive. Int J Biol Macromol 2019; 132: 1121-30.
       [http://dx.doi.org/10.1016/j.ijbiomac.2019.03.166] [PMID: 30928374]

[27]   Wang M, Du Y, Huang H, *et al.* Silk fibroin peptide suppresses proliferation and induces apoptosis and cell cycle arrest in human lung cancer cells. Acta Pharmacol Sin 2019; 40(4): 522-9.
       [http://dx.doi.org/10.1038/s41401-018-0048-0] [PMID: 29921888]

[28]   Khodarahmi Borujeni E, Shaikhzadeh Najar S, Kamali Dolatabadi M. The study on structural properties and tensile strength of reared silkworm cocoon. J Textil Inst 2018; 109(2): 195-201.
       [http://dx.doi.org/10.1080/00405000.2017.1335378]

[29]   Wang Y, Cai R, Tao G, *et al.* A Novel AgNPs/sericin/agar film with enhanced mechanical property and antibacterial capability. Molecules 2018; 23(7): 1821.
       [http://dx.doi.org/10.3390/molecules23071821] [PMID: 30041405]

[30]   Yu K, Lu F, Li Q, *et al. In situ* assembly of Ag nanoparticles (AgNPs) on porous silkworm cocoon-based wound film: enhanced antimicrobial and wound healing activity. Sci Rep 2017; 7(1): 2107.
       [http://dx.doi.org/10.1038/s41598-017-02270-6] [PMID: 28522813]

[31]   He H, Tao G, Wang Y, *et al. In situ* green synthesis and characterization of sericin-silver nanoparticle composite with effective antibacterial activity and good biocompatibility. Mater Sci Eng C 2017; 80: 509-16.
[http://dx.doi.org/10.1016/j.msec.2017.06.015] [PMID: 28866194]

[32]   Yu K, Lu F, Li Q, *et al.* Accelerated wound-healing capabilities of a dressing fabricated from silkworm cocoon. Int J Biol Macromol 2017; 102: 901-13.
[http://dx.doi.org/10.1016/j.ijbiomac.2017.04.069] [PMID: 28435057]

[33]   Züge LCB, Silva VR, Hamerski F, Ribani M, Gimenes ML, Scheer AP. Emulsifying properties of sericin obtained from hot water degumming process. J Food Process Eng 2017; 40(1): e12267.
[http://dx.doi.org/10.1111/jfpe.12267]

[34]   Guan J, Zhu W, Liu B, Yang K, Vollrath F, Xu J. Comparing the microstructure and mechanical properties of *Bombyx mori* and *Antheraea pernyi* cocoon composites. Acta Biomater 2017; 47: 60-70.
[http://dx.doi.org/10.1016/j.actbio.2016.09.042] [PMID: 27693687]

[35]   Liu L, Wang J, Duan S, *et al.* Systematic evaluation of sericin protein as a substitute for fetal bovine serum in cell culture. Sci Rep 2016; 6(1): 31516.
[http://dx.doi.org/10.1038/srep31516] [PMID: 27531556]

[36]   Guo X, Dong Z, Zhang Y, Li Y, Liu H, Xia Q, *et al.* Proteins in the cocoon of silkworm inhibit the growth of *Beauveria bassiana.* PLoS ONE 2016; 11(3): e0151764.

[37]   Vidart J, Nakashima M, da Silva T, Rosa P, Gimenes M, Vieira M, *et al.* Sericin and alginate blend as matrix for incorporation of diclofenac sodium. Chem Eng Trans 2016; 52: 343-8.

[38]   Siritientong T, Aramwit P. Characteristics of carboxymethyl cellulose/sericin hydrogels and the influence of molecular weight of carboxymethyl cellulose. Macromol Res 2015; 23(9): 861-6.
[http://dx.doi.org/10.1007/s13233-015-3116-z]

[39]   Aramwit P, Palapinyo S, Srichana T, Chottanapund S, Muangman P. Silk sericin ameliorates wound healing and its clinical efficacy in burn wounds. Arch Dermatol Res 2013; 305(7): 585-94.
[http://dx.doi.org/10.1007/s00403-013-1371-4] [PMID: 23748948]

[40]   Yun H, Oh H, Kim MK, *et al.* Extraction conditions of Antheraea mylitta sericin with high yields and minimum molecular weight degradation. Int J Biol Macromol 2013; 52: 59-65.
[http://dx.doi.org/10.1016/j.ijbiomac.2012.09.017] [PMID: 23026092]

[41]   Kaewkorn W, Limpeanchob N, Tiyaboonchai W, Pongcharoen S, Sutheerawattananonda M. Effects of silk sericin on the proliferation and apoptosis of colon cancer cells. Biol Res 2012; 45(1): 45-50.
[http://dx.doi.org/10.4067/S0716-97602012000100006] [PMID: 22688983]

[42]   Oh H, Kim MK, Lee KH. Preparation of sericin microparticles by electrohydrodynamic spraying and their application in drug delivery. Macromol Res 2011; 19(3): 266-72.
[http://dx.doi.org/10.1007/s13233-011-0301-6]

[43]   Zhang H, Yang M, Min S, Feng Q, Gao X, Zhu L. Preparation and characterization of a novel spongy hydrogel from aqueous *Bombyx mori* sericin. E-Polym 2008; 8(1)

[44]   Jiang P, Liu H, Wang C, Wu L, Huang J, Guo C. Tensile behavior and morphology of differently degummed silkworm (*Bombyx mori*) cocoon silk fibres. Mater Lett 2006; 60(7): 919-25.
[http://dx.doi.org/10.1016/j.matlet.2005.10.056]

[45]   Terada S, Nishimura T, Sasaki M, Yamada H, Miki M. Sericin, a protein derived from silkworms, accelerates the proliferation of several mammalian cell lines including a hybridoma. Cytotechnology 2002; 40(1/3): 3-12.
[http://dx.doi.org/10.1023/A:1023993400608] [PMID: 19003099]

[46]   Ontong JC, Singh S, Nwabor OF, Chusri S, Voravuthikunchai SP. Potential of antimicrobial topical gel with synthesized biogenic silver nanoparticle using *Rhodomyrtus tomentosa* leaf extract and silk sericin. Biotechnol Lett 2020; 42(12): 2653-64.

[http://dx.doi.org/10.1007/s10529-020-02971-5] [PMID: 32683522]

[47]   Silva T, Silva A, Martins J, Vieira M, Gimenes M, Silva M. Evaluation of drug delivery of diclofenac sodium in simulated gastric and enteric systems by mucoadhesive sericin-alginate particles. Chem Eng Trans 2015; 43: 823-8.

[48]   Verdanova M, Pytlik R, Kalbacova MH. Evaluation of sericin as a fetal bovine serum-replacing cryoprotectant during freezing of human mesenchymal stromal cells and human osteoblast-like cells. Biopreserv Biobank 2014; 12(2): 99-105.
[http://dx.doi.org/10.1089/bio.2013.0078] [PMID: 24749876]

[49]   Syukri DM, Nwabor OF, Singh S, *et al.* Antibacterial-coated silk surgical sutures by *ex situ* deposition of silver nanoparticles synthesized with *Eucalyptus camaldulensis* eradicates infections. J Microbiol Methods 2020; 174: 105955.
[http://dx.doi.org/10.1016/j.mimet.2020.105955] [PMID: 32442657]

[50]   Pattabiraman VR, Bode JW. Rethinking amide bond synthesis. Nature 2011; 480(7378): 471-9.
[http://dx.doi.org/10.1038/nature10702] [PMID: 22193101]

[51]   Sánchez A, Vázquez A. Bioactive peptides: A review. Food Quality and Safety 2017; 1(1): 29-46.
[http://dx.doi.org/10.1093/fqs/fyx006]

[52]   Peng L, Kong X, Wang Z, Ai-lati A, Ji Z, Mao J. Baijiu vinasse as a new source of bioactive peptides with antioxidant and anti-inflammatory activity. Food Chem 2021; 339: 128159.
[http://dx.doi.org/10.1016/j.foodchem.2020.128159] [PMID: 33152898]

[53]   Haque E, Chand R. Antihypertensive and antimicrobial bioactive peptides from milk proteins. Eur Food Res Technol 2008; 227(1): 7-15.
[http://dx.doi.org/10.1007/s00217-007-0689-6]

[54]   Skjånes K, Aesoy R, Herfindal L, Skomedal H. Bioactive peptides from microalgae: Focus on anti-cancer and immunomodulating activity. Physiol Plant 2021; 173(2): 612-23.
[http://dx.doi.org/10.1111/ppl.13472] [PMID: 34085279]

[55]   Mahdi C, Untari H, Padaga MC, Eds. Identification and characterization of bioactive peptides of fermented goat milk as a sources of antioxidant as a therapeutic natural product. IOP Conference Series Mater Sci Eng.
[http://dx.doi.org/10.1088/1757-899X/299/1/012014]

[56]   Erdmann K, Cheung BWY, Schröder H. The possible roles of food-derived bioactive peptides in reducing the risk of cardiovascular disease. J Nutr Biochem 2008; 19(10): 643-54.
[http://dx.doi.org/10.1016/j.jnutbio.2007.11.010] [PMID: 18495464]

[57]   Briuglia ML, Urquhart AJ, Lamprou DA. Sustained and controlled release of lipophilic drugs from a self-assembling amphiphilic peptide hydrogel. Int J Pharm 2014; 474(1-2): 103-11.
[http://dx.doi.org/10.1016/j.ijpharm.2014.08.025] [PMID: 25148727]

[58]   Nagai Y, Unsworth LD, Koutsopoulos S, Zhang S. Slow release of molecules in self-assembling peptide nanofiber scaffold. J Control Release 2006; 115(1): 18-25.
[http://dx.doi.org/10.1016/j.jconrel.2006.06.031] [PMID: 16962196]

[59]   Kumada Y, Hammond NA, Zhang S. Functionalized scaffolds of shorter self-assembling peptides containing MMP-2 cleavable motif promote fibroblast proliferation and significantly accelerate 3-D cell migration independent of scaffold stiffness. Soft Matter 2010; 6(20): 5073-9.
[http://dx.doi.org/10.1039/c0sm00333f]

[60]   Jung JP, Jones JL, Cronier SA, Collier JH. Modulating the mechanical properties of self-assembled peptide hydrogels *via* native chemical ligation. Biomaterials 2008; 29(13): 2143-51.
[http://dx.doi.org/10.1016/j.biomaterials.2008.01.008] [PMID: 18261790]

[61]   Altunbas A, Lee SJ, Rajasekaran SA, Schneider JP, Pochan DJ. Encapsulation of curcumin in self-assembling peptide hydrogels as injectable drug delivery vehicles. Biomaterials 2011; 32(25): 5906-14.

[http://dx.doi.org/10.1016/j.biomaterials.2011.04.069] [PMID: 21601921]

[62] Jabbari E, Yang X, Moeinzadeh S, He X. Drug release kinetics, cell uptake, and tumor toxicity of hybrid VVVVVVKK peptide-assembled polylactide nanoparticles. Eur J Pharm Biopharm 2013; 84(1): 49-62.
[http://dx.doi.org/10.1016/j.ejpb.2012.12.012] [PMID: 23275111]

[63] Murphy EA, Majeti BK, Barnes LA, *et al.* Nanoparticle-mediated drug delivery to tumor vasculature suppresses metastasis. Proc Natl Acad Sci USA 2008; 105(27): 9343-8.
[http://dx.doi.org/10.1073/pnas.0803728105] [PMID: 18607000]

[64] Li L, Xiang D, Shigdar S, *et al.* Epithelial cell adhesion molecule aptamer functionalized PLGA-lecithin-curcumin-PEG nanoparticles for targeted drug delivery to human colorectal adenocarcinoma cells. Int J Nanomedicine 2014; 9: 1083-96.
[PMID: 24591829]

[65] Aravind A, Jeyamohan P, Nair R, *et al.* AS1411 aptamer tagged PLGA-lecithin-PEG nanoparticles for tumor cell targeting and drug delivery. Biotechnol Bioeng 2012; 109(11): 2920-31.
[http://dx.doi.org/10.1002/bit.24558] [PMID: 22615073]

[66] Chen Y, Wang J, Wang J, *et al.* Aptamer functionalized cisplatin-albumin nanoparticles for targeted delivery to epidermal growth factor receptor positive cervical cancer. J Biomed Nanotechnol 2016; 12(4): 656-66.
[http://dx.doi.org/10.1166/jbn.2016.2203] [PMID: 27301192]

[67] Xu L, He XY, Liu BY, *et al.* Aptamer-functionalized albumin-based nanoparticles for targeted drug delivery. Colloids Surf B Biointerfaces 2018; 171: 24-30.
[http://dx.doi.org/10.1016/j.colsurfb.2018.07.008] [PMID: 30005287]

[68] Saleh T, Soudi T, Shojaosadati SA. Aptamer functionalized curcumin-loaded human serum albumin (HSA) nanoparticles for targeted delivery to HER-2 positive breast cancer cells. Int J Biol Macromol 2019; 130: 109-16.
[http://dx.doi.org/10.1016/j.ijbiomac.2019.02.129] [PMID: 30802519]

[69] Rosillo-Calle F, Woods J. The biomass assessment handbook. Routledge 2012.
[http://dx.doi.org/10.4324/9781849772884]

[70] Bajwa DS, Pourhashem G, Ullah AH, Bajwa SG. A concise review of current lignin production, applications, products and their environmental impact. Ind Crops Prod 2019; 139: 111526.
[http://dx.doi.org/10.1016/j.indcrop.2019.111526]

[71] Policy USEPAOo. Inventory of US greenhouse gas emissions and sinks, 1990-1994. US Environmental Protection Agency, Office of Policy, Planning, and Evaluation 1995.

[72] De Candolle AP, De Candolle A. Theorie elementaire de la botanique, ou, Exposition des principes de la classification naturelle et de l'art de décrire et d'étudier les végétaux: Roret. 1844.

[73] Baker DA, Gallego NC, Baker FS. On the characterization and spinning of an organic-purified lignin toward the manufacture of low-cost carbon fiber. J Appl Polym Sci 2012; 124(1): 227-34.
[http://dx.doi.org/10.1002/app.33596]

[74] Tribot A, Amer G, Abdou Alio M, *et al.* Wood-lignin: Supply, extraction processes and use as bio-based material. Eur Polym J 2019; 112: 228-40.
[http://dx.doi.org/10.1016/j.eurpolymj.2019.01.007]

[75] Latham KG, Matsakas L, Figueira J, Rova U, Christakopoulos P, Jansson S. Examination of how variations in lignin properties from Kraft and organosolv extraction influence the physicochemical characteristics of hydrothermal carbon. J Anal Appl Pyrolysis 2021; 155: 105095.
[http://dx.doi.org/10.1016/j.jaap.2021.105095]

[76] Jacob John M, Thomas S. Natural polymers: an overview. Natural polymers, The Royal Society of Chemistry. 2012; 1: pp. (1)1-7.

[77]   Zhang J, Almoallim HS, Ali Alharbi S, Yang B. Anti-atherosclerotic activity of Betulinic acid loaded polyvinyl alcohol/methylacrylate grafted Lignin polymer in high fat diet induced atherosclerosis model rats. Arab J Chem 2021; 14(2): 102934.
        [http://dx.doi.org/10 1016/j.arabjc.2020.102934]

[78]   Lee SJ, Begildayeva T, Yeon S, *et al.* Eco-friendly synthesis of lignin mediated silver nanoparticles as a selective sensor and their catalytic removal of aromatic toxic nitro compounds. Environ Pollut 2021; 269: 116174.
        [http://dx.doi.org/10 1016/j.envpol.2020.116174] [PMID: 33280906]

[79]   Mishra PK, Wimmer R. Aerosol assisted self-assembly as a route to synthesize solid and hollow spherical lignin colloids and its utilization in layer by layer deposition. Ultrason Sonochem 2017; 35(Pt A): 45-50.
        [http://dx.doi.org/10.1016/j.ultsonch.2016.09.001] [PMID: 27614582]

[80]   Tardy BL, Richardson JJ, Guo J, Lehtonen J, Ago M, Rojas OJ. Lignin nano- and microparticles as template for nanostructured materials: formation of hollow metal-phenolic capsules. Green Chem 2018; 20(6): 1335-44.
        [http://dx.doi.org/10.1039/C8GC00064F]

[81]   Farooq M, Zou T, Riviere G, Sipponen MH, Österberg M. Strong, Ductile, and waterproof cellulose nanofibril composite films with colloidal lignin particles. Biomacromolecules 2019; 20(2): 693-704.
        [http://dx.doi.org/10.1021/acs.biomac.8b01364] [PMID: 30358992]

[82]   Nakasone K, Kobayashi T. Cytocompatible cellulose hydrogels containing trace lignin. Mater Sci Eng C 2016; 64: 269-77.
        [http://dx.doi.org/10.1016/j.msec.2016.03.108] [PMID: 27127053]

[83]   Pathayappurakkal Mohanan D, Pathayappurakkal Mohan N, Selvasudha N, Thekkilaveedu S, Kandasamy R. Facile fabrication and structural elucidation of lignin based macromolecular green composites for multifunctional applications. J Appl Polym Sci 2021; 138(43): 51280.
        [http://dx.doi.org/10.1002/app.51280]

[84]   Panariello L, Vannozzi A, Morganti P, Coltelli MB, Lazzeri A. Biobased and eco-compatible beauty films coated with chitin nanofibrils, nanolignin and vitamin E. Cosmetics 2021; 8(2): 27.
        [http://dx.doi.org/10.3390/cosmetics8020027]

[85]   Fukuda N, Hatakeyama M, Kitaoka T. Enzymatic Preparation and characterization of spherical microparticles composed of artificial lignin and tempo-oxidized cellulose nanofiber. Nanomaterials (Basel) 2021; 11(4): 917.
        [http://dx.doi.org/10.3390/nano11040917] [PMID: 33916825]

[86]   Rahdar A, Sargazi S, Barani M, Shahraki S, Sabir F, Aboudzadeh M. Lignin-stabilized doxorubicin microemulsions: synthesis, physical characterization, and *in vitro* assessments. Polymers (Basel) 2021; 13(4): 641.
        [http://dx.doi.org/10.3390/polym13040641] [PMID: 33670009]

[87]   Akbari S, Bahi A, Farahani A, Milani AS, Ko F. Fabrication and characterization of lignin/dendrimer electrospun blended fiber mats. Molecules 2021; 26(3): 518.
        [http://dx.doi.org/10.3390/molecules26030518] [PMID: 33498227]

[88]   Abudula T, Gauthaman K, Mostafavi A, *et al.* Sustainable drug release from polycaprolactone coated chitin-lignin gel fibrous scaffolds. Sci Rep 2020; 10(1): 20428.
        [http://dx.doi.org/10.1038/s41598-020-76971-w] [PMID: 33235239]

[89]   Morena AG, Stefanov I, Ivanova K, Pérez-Rafael S, Sánchez-Soto M, Tzanov T. Antibacterial polyurethane foams with incorporated lignin-capped silver nanoparticles for chronic wound treatment. Ind Eng Chem Res 2020; 59(10): 4504-14.
        [http://dx.doi.org/10.1021/acs.iecr.9b06362]

[90]   Busatto CA, Taverna ME, Lescano MR, Zalazar C, Estenoz DA. Preparation and characterization of

lignin microparticles-in-alginate beads for atrazine controlled release. J Polym Environ 2019; 27(12): 2831-41.
[http://dx.doi.org/10.1007/s10924-019-01564-2]

[91] Domínguez-Robles J, Stewart SA, Rendl A, González Z, Donnelly RF, Larrañeta E. Lignin and cellulose blends as pharmaceutical excipient for tablet manufacturing *via* direct compression. Biomolecules 2019; 9(9): 423.
[http://dx.doi.org/10.3390/biom9090423] [PMID: 31466387]

[92] Belgodere JA, Zamin SA, Kalinoski RM, *et al.* Modulating Mechanical Properties of Collagen–Lignin Composites. ACS Appl Bio Mater 2019; 2(8): 3562-72.
[http://dx.doi.org/10.1021/acsabm.9b00444] [PMID: 35030742]

[93] A. zghair AL-Suwaytee E, A. Al-Mayyahi B, J.M. ALFartosy A. Study of thermal behavior and anti-breast cancer activity of some new lignin-nanoparticle networks sustained with triterpenoid compound isolated from *Calotropis procera* L. leaves. Egypt J Chem 2019; 62(4): 583-608.

[94] Pishnamazi M, Casilagan S, Clancy C, *et al.* Microcrystalline cellulose, lactose and lignin blends: Process mapping of dry granulation *via* roll compaction. Powder Technol 2019; 341: 38-50.
[http://dx.doi.org/10.1016/j.powtec.2018.07.003]

[95] Mattinen ML, Riviere G, Henn A, *et al.* Colloidal lignin particles as adhesives for soft materials. Nanomaterials (Basel) 2018; 8(12): 1001.
[http://dx.doi.org/10.3390/nano8121001] [PMID: 30513957]

[96] Zmejkoski D, Spasojević D, Orlovska I, *et al.* Bacterial cellulose-lignin composite hydrogel as a promising agent in chronic wound healing. Int J Biol Macromol 2018; 118(Pt A): 494-503.
[http://dx.doi.org/10.1016/j.ijbiomac.2018.06.067] [PMID: 29909035]

[97] Dai L, Zhu W, Liu R, Si C. Lignin-containing self-nanoemulsifying drug delivery system for enhance stability and oral absorption of trans-resveratrol. Part Part Syst Charact 2018; 35(4): 1700447.
[http://dx.doi.org/10.1002/ppsc.201700447]

[98] Yang W, Fortunati E, Gao D, *et al.* Valorization of acid isolated high yield lignin nanoparticles as innovative antioxidant/antimicrobial organic materials. ACS Sustain Chem& Eng 2018; 6(3): 3502-14.
[http://dx.doi.org/10.1021/acssuschemeng.7b03782]

[99] Reesi F, Minaiyan M, Taheri A. A novel lignin-based nanofibrous dressing containing arginine for wound-healing applications. Drug Deliv Transl Res 2018; 8(1): 111-22.
[http://dx.doi.org/10.1007/s13346-017-0441-0] [PMID: 29159695]

[100] Figueiredo P, Lintinen K, Kiriazis A, *et al. In vitro* evaluation of biodegradable lignin-based nanoparticles for drug delivery and enhanced antiproliferation effect in cancer cells. Biomaterials 2017; 121: 97-108.
[http://dx.doi.org/10.1016/j.biomaterials.2016.12.034] [PMID: 28081462]

[101] Frangville C, Rutkevičius M, Richter AP, Velev OD, Stoyanov SD, Paunov VN. Fabrication of environmentally biodegradable lignin nanoparticles. ChemPhysChem 2012; 13(18): 4235-43.
[http://dx.doi.org/10.1002/cphc.201200537] [PMID: 23047584]

[102] Lu Q, Zhu M, Zu Y, *et al.* Comparative antioxidant activity of nanoscale lignin prepared by a supercritical antisolvent (SAS) process with non-nanoscale lignin. Food Chem 2012; 135(1): 63-7.
[http://dx.doi.org/10.1016/j.foodchem.2012.04.070]

[103] Çalgeris İ, Çakmakçı E, Ogan A, Kahraman MV, Kayaman-Apohan N. Preparation and drug release properties of lignin-starch biodegradable films. Stärke 2012; 64(5): 399-407.
[http://dx.doi.org/10.1002/star.201100158]

[104] Ciolacu D, Oprea AM, Anghel N, Cazacu G, Cazacu M. New cellulose–lignin hydrogels and their application in controlled release of polyphenols. Mater Sci Eng C 2012; 32(3): 452-63.
[http://dx.doi.org/10.1016/j.msec.2011.11.018]

[105] Flores-Céspedes F, Figueredo-Flores CI, Daza-Fernández I, Vidal-Peña F, Villafranca-Sánchez M,

Fernández-Pérez M. Preparation and characterization of imidacloprid lignin-polyethylene glycol matrices coated with ethylcellulose. J Agric Food Chem 2012; 60(4): 1042-51.
[http://dx.doi.org/10.1021/jf2037483] [PMID: 22224401]

[106] Chen X, Li Z, Zhang L, *et al.* Preparation of a novel lignin-based film with high solid content and its physicochemical characteristics. Ind Crops Prod 2021; 164: 113396.
[http://dx.doi.org/10.1016/j.indcrop.2021.113396]

[107] Cao Q, Wu Q, Dai L, *et al.* Size-controlled lignin nanoparticles for tuning the mechanical properties of poly(vinyl alcohol). Ind Crops Prod 2021; 172: 114012.
[http://dx.doi.org/10.1016/j.indcrop.2021.114012]

[108] Yu Y, Naik SS, Oh Y, Theerthagiri J, Lee SJ, Choi MY. Lignin-mediated green synthesis of functionalized gold nanoparticles *via* pulsed laser technique for selective colorimetric detection of lead ions in aqueous media. J Hazard Mater 2021; 420: 126585.
[http://dx.doi.org/10.1016/j.jhazmat.2021.126585] [PMID: 34273885]

[109] Slavin YN, Ivanova K, Hoyo J, *et al.* Novel lignin-capped silver nanoparticles against multidrug-resistant bacteria. ACS Appl Mater Interfaces 2021; 13(19): 22098-109.
[http://dx.doi.org/10.1021/acsami.0c16921] [PMID: 33945683]

[110] Petrie FA, Gorham JM, Busch RT, Leontsev SO, Ureña-Benavides EE, Vasquez ES. Facile fabrication and characterization of kraft lignin@Fe_3O_4 nanocomposites using pH driven precipitation: Effects on increasing lignin content. Int J Biol Macromol 2021; 181: 313-21.
[http://dx.doi.org/10.1016/j.ijbiomac.2021.03.105] [PMID: 33766601]

[111] Wasti S, Triggs E, Farag R, *et al.* Influence of plasticizers on thermal and mechanical properties of biocomposite filaments made from lignin and polylactic acid for 3D printing. Compos, Part B Eng 2021; 205: 108483.
[http://dx.doi.org/10.1016/j.compositesb.2020.108483]

[112] Chen X, Xi X, Pizzi A, *et al.* Oxidized demethylated lignin as a bio-based adhesive for wood bonding. J Adhes 2021; 97(9): 873-90.
[http://dx.doi.org/10.1080/00218464.2019.1710830]

[113] Alwadani N, Ghavidel N, Fatehi P. Surface and interface characteristics of hydrophobic lignin derivatives in solvents and films. Colloids Surf A Physicochem Eng Asp 2021; 609: 125656.
[http://dx.doi.org/10.1016/j.colsurfa.2020.125656]

[114] Zhen X, Li H, Xu Z, *et al.* Facile synthesis of lignin-based epoxy resins with excellent thermal-mechanical performance. Int J Biol Macromol 2021; 182: 276-85.
[http://dx.doi.org/10.1016/j.ijbiomac.2021.03.203] [PMID: 33838187]

[115] Shirwaikar AA, Joseph A, Srinivasan KK, Jacob S. Novel co-processed excipients of mannitol and microcrystalline cellulose for preparing fast dissolving tablets of glipizide. Indian J Pharm Sci 2007; 69(5): 633.
[http://dx.doi.org/10.4103/0250-474X.38467]

[116] Tsolaki E, Stocker MW, Healy AM, Ferguson S. Formulation of ionic liquid APIs *via* spray drying processes to enable conversion into single and two-phase solid forms. Int J Pharm 2021; 603: 120669.
[http://dx.doi.org/10.1016/j.ijpharm.2021.120669] [PMID: 33989753]

[117] Ermawati DE, Andini BP, Prihapsara F, *et al.* Optimization of Suweg starch (*Amorphophallus paeoniifolius* (Dennst.) Nicolson) and lactose as co-processed excipient of Ibuprofen-PEG 6000 solid dispersion of effervescent tablet. AIP Conf Proc 2020; 2237(1): 020061.
[http://dx.doi.org/10.1063/5.0005632]

[118] Singh S, Nwabor OF, Ontong JC, Voravuthikunchai SP. Characterization and assessment of compression and compactibility of novel spray-dried, co-processed bio-based polymer. J Drug Deliv Sci Technol 2020; 56: 101526.
[http://dx.doi.org/10.1016/j.jddst.2020.101526]

[119]   Singh S, Nwabor OF, Ontong JC, Kaewnopparat N, Voravuthikunchai SP. Characterization of a novel, co-processed bio-based polymer, and its effect on mucoadhesive strength. Int J Biol Macromol 2020; 145: 865-75.
[http://dx.doi.org/10.1016/j.ijbiomac.2019.11.198] [PMID: 31783076]

[120]   Mori D, Rathod P, Parmar R, Dudhat K, Chavda J. Preparation and optimization of multi-functional directly compressible excipient: an integrated approach of principal component analysis and design of experiments. Drug Dev Ind Pharm 2020; 46(12): 2010-21.
[http://dx.doi.org/10.1080/03639045.2020.1841788] [PMID: 33095675]

[121]   Patel J, Mori D. Application of 3^2 full factorial design and desirability function for optimizing the manufacturing process for directly compressible multi-functional co-processed excipient. Curr Drug Deliv 2020; 17(6): 523-39.
[http://dx.doi.org/10.2174/1567201817666200508094743] [PMID: 32384027]

[122]   Bernal Rodriguez CA, Bassani VL, Castellanos L, Ramos Rodríguez FA, Baena Y. Development of an oral control release system from *Physalis peruvia*na L. fruits extract based on the co-spray-drying method. Powder Technol 2019; 354: 676-88.
[http://dx.doi.org/10.1016/j.powtec.2019.06.024]

[123]   Assaf SM, Subhi Khanfar M, Bassam Farhan A, Said Rashid I, Badwan AA. Preparation and characterization of co-processed starch/MCC/chitin hydrophilic polymers onto magnesium *silica*te. Pharm Dev Technol 2019; 24(6): 761-74.
[http://dx.doi.org/10.1080/10837450.2019.1596131] [PMID: 30888873]

[124]   Apeji YE, Olayemi OJ, Anyebe SN, *et al.* Impact of binder as a formulation variable on the material and tableting properties of developed co-processed excipients. SN Applied Sciences 2019; 1(6): 561.
[http://dx.doi.org/10.1007/s42452-019-0585-2]

[125]   Patki M, Patel K. Development of a solid supersaturated self-nanoemulsifying preconcentrate (S-superSNEP) of fenofibrate using dimethylacetamide and a novel co-processed excipient. Drug Dev Ind Pharm 2019; 45(3): 405-14.
[http://dx.doi.org/10.1080/03639045.2018.1546311] [PMID: 30444435]

[126]   Castañeda Hernández O. Oswaldo castañeda hernández Ehb, Enrique amador gonzález, Luz maría melgoza contreras. Production of directly compressible excipients with mannitol by wet granulation: rheological, compressibility and compactibility characterization. Farmacia 2019; 67(6): 973-85.
[http://dx.doi.org/10.31925/farmacia.2019.6.7]

[127]   Olorunsola EO, Usungurua SG. Effect of polymer ratio on the quality of a co-processed excipient of prosopis gum and crab shell chitosan. Trop J Pharm Res 2018; 17(9): 1693-9.
[http://dx.doi.org/10.4314/tjpr.v17i9.2]

[128]   Patel S, Patel N, Misra M, Joshi A. Controlled-release domperidone pellets compressed into fast disintegrating tablets forming a multiple-unit pellet system (MUPS). J Drug Deliv Sci Technol 2018; 45: 220-9.
[http://dx.doi.org/10.1016/j.jddst.2017.12.015]

[129]   Olufunke D. Akin-Ajani TOA, Uchenna M. Okoli, Ozioma Okonta Development of directly compressible excipients from *Phoenix dactylifera* (Date) mucilage and microcrystalline cellulose using co-processing techniques. ACTA Pharm Sci 2018; 56(3): 7-25.

[130]   Pattnaik S, Swain K, Rao JV, Talla V, Prusty KB, Subudhi SK. Polymer co-processing of ibuprofen through compaction for improved oral absorption. RSC Advances 2015; 5(91): 74720-5.
[http://dx.doi.org/10.1039/C5RA13038G]

[131]   Sharma P, Modi SR, Bansal AK. Co-processing of hydroxypropyl methylcellulose (HPMC) for improved aqueous dispersibility. Int J Pharm 2015; 485(1-2): 348-56.
[http://dx.doi.org/10.1016/j.ijpharm.2015.03.036] [PMID: 25796127]

[132]   Goyanes A, Martínez-Pacheco R. New co-processed MCC-based excipient for fast release of low

solubility drugs from pellets prepared by extrusion-spheronization. Drug Dev Ind Pharm 2015; 41(3): 362-8.
[http://dx.doi.org/10.3109/03639045.2013.861479] [PMID: 24279425]

[133] Gallo L, Piña J, Bucalá V, Allemandi D, Ramírez-Rigo MV. Development of a modified-release hydrophilic matrix system of a plant extract based on co-spray-dried powders. Powder Technol 2013; 241: 252-62.
[http://dx.doi.org/10.1016/j.powtec.2013.03.011]

[134] Lin X, Chyi CW, Ruan K, Feng Y, Heng PWS. Development of potential novel cushioning agents for the compaction of coated multi-particulates by co-processing micronized lactose with polymers. Eur J Pharm Biopharm 2011; 79(2): 406-15.
[http://dx.doi.org/10 1016/j.ejpb.2011.03.024] [PMID: 21458566]

[135] Coucke D, Vervaet C, Foreman P, Adriaensens P, Carleer R, Remon JP. Effect on the nasal bioavailability of co-processing drug and bioadhesive carrier *via* spray-drying. Int J Pharm 2009; 379(1): 67-71.
[http://dx.doi.org/10 1016/j.ijpharm.2009.06.008] [PMID: 19539738]

[136] Tranová T, Macho O, Loskot J, Mužíková J. Study of rheological and tableting properties of lubricated mixtures of co-processed dry binders for orally disintegrating tablets. Eur J Pharm Sci 2022; 168: 106035.
[http://dx.doi.org/10.1016/j.ejps.2021.106035] [PMID: 34634469]

[137] Mahto A, Mishra S. Design, development and validation of guar gum based pH sensitive drug delivery carrier *via* graft copolymerization reaction using microwave irradiations. Int J Biol Macromol 2019; 138: 278-91.
[http://dx.doi.org/10.1016/j.ijbiomac.2019.07.063] [PMID: 31310787]

[138] Roland C. Interpenetrating Polymer Networks (IPN Q1). Structure and Mechanical Behavior. Encylop Polym Nanomat 2013; pp. 1-10.

[139] Roy D, Semsarilar M, Guthrie JT, Perrier S. Cellulose modification by polymer grafting: a review. Chem Soc Rev 2009; 38(7): 2046-64.
[http://dx.doi.org/10.1039/b808639g] [PMID: 19551181]

[140] Bhattacharya Amit PR. Basic feature and techniques. In: Amit Bhattacharya, Paramita Ray, Eds. Polymer grafting and crosslinking: John Wiley and Sons. 2008; p. 8.
[http://dx.doi.org/10.1002/9780470414811.ch2]

[141] Mate CJ, Mishra S, Srivastava PK. *In vitro* release kinetics of graft matrices from Lannea coromandelica (Houtt) gum for treatment of colonic diseases by 5-ASA. Int J Biol Macromol 2020; 149: 908-20.
[http://dx.doi.org/10.1016/j.ijbiomac.2020.02.009] [PMID: 32027894]

[142] Singh A, Mangla B. Sethi S, Kamboj S, Sharma R, Rana V. QbD based synthesis and characterization of polyacrylamide grafted corn fibre gum. Carbohydr Polym 2017; 156: 45-55.
[http://dx.doi.org/10.1016/j.carbpol.2016.08.089] [PMID: 27842845]

[143] Bidarakatte Krishnappa P, Badalamoole V. Karaya gum-graft-poly(2-(dimethylamino)ethyl methacrylate) gel: An efficient adsorbent for removal of ionic dyes from water. Int J Biol Macromol 2019; 122: 997-1007.
[http://dx.doi.org/10.1016/j.ijbiomac.2018.09.038] [PMID: 30201563]

[144] Guleria A, Kumari G, Lima EC. Cellulose-g-poly-(acrylamide-co-acrylic acid) polymeric bioadsorbent for the removal of toxic inorganic pollutants from wastewaters. Carbohydr Polym 2020; 228: 115396.
[http://dx.doi.org/10.1016/j.carbpol.2019.115396] [PMID: 31635743]

[145] Beyaz K, Vaca-Garcia C, Vedrenne E, Haddadine N, Benaboura A, Thiebaud-Roux S. Graft copolymerization of hydroxyethyl cellulose with solketal acrylate: Preparation and characterization for moisture absorption application. IJPAC Int J Polym Anal Charact 2019; 24(3): 245-56.
[http://dx.doi.org/10.1080/1023666X.2019.1567085]

[146]   Weerapoprasit C, Prachayawarakorn J. Characterization and properties of biodegradable thermoplastic grafted starch films by different contents of methacrylic acid. Int J Biol Macromol 2019; 123: 657-63.
[http://dx.doi.org/10.1016/j.ijbiomac.2018.11.083] [PMID: 30445086]

[147]   Czarnecka E, Nowaczyk J. Semi-natural superabsorbents based on starch-g-poly(acrylic acid): modification, synthesis and application. Polymers (Basel) 2020; 12(8): 1794.
[http://dx.doi.org/10.3390/polym12081794] [PMID: 32785178]

[148]   Patra S, Bala NN, Nandi G. Synthesis, characterization and fabrication of sodium carboxymethyl-okr--gum-grafted-polymethacrylamide into sustained release tablet matrix. Int J Biol Macromol 2020; 164: 3885-900.
[http://dx.doi.org/10.1016/j.ijbiomac.2020.09.025] [PMID: 32910964]

[149]   Ghapar NFA, Baharin H, Karim KJA. Preparation and characterization of chitosan beads grafted with poly(methyl methacrylate) for controlled release study. AIP Conf Proc 2018; 1985(1): 050001.
[http://dx.doi.org/10.1063/1.5047195]

[150]   Alfaifi AYA, El-Newehy MH, Abdel-Halim ES, Al-Deyab SS. Microwave-assisted graft copolymerization of amino acid based monomers onto starch and their use as drug carriers. Carbohydr Polym 2014; 106: 440-52.
[http://dx.doi.org/10.1016/j.carbpol.2014.01.028] [PMID: 24721100]

[151]   Kumar A, De A, Mozumdar S. Synthesis of acrylate guar-gum for delivery of bio-active molecules. Bull Mater Sci 2015; 38(4): 1025-32.
[http://dx.doi.org/10.1007/s12034-015-0930-z]

[152]   Credou J, Faddoul R, Berthelot T. One-step and eco-friendly modification of cellulose membranes by polymer grafting. RSC Advances 2014; 4(105): 60959-69.
[http://dx.doi.org/10.1039/C4RA11219A]

[153]   Bardajee GR, Hooshyar Z, Kabiri F. Preparation and investigation on swelling and drug delivery properties of a novel silver/salep-g-poly(acrylic Acid) nanocomposite hydrogel. Bull Korean Chem Soc 2012; 33(8): 2635-41.
[http://dx.doi.org/10.5012/bkcs.2012.33.8.2635]

[154]   Cao L, Wang X, Wang G, Wang J. A pH-sensitive porous chitosan membrane prepared *via* surface grafting copolymerization in supercritical carbon dioxide. Polym Int 2015; 64(3): 383-8.
[http://dx.doi.org/10.1002/pi.4798]

[155]   Zhang J, Xiao H, Li N, Ping Q, Zhang Y. Synthesis and characterization of super-absorbent hydrogels based on hemicellulose. J Appl Polym Sci 2015; 132(34): n/a.
[http://dx.doi.org/10.1002/app.42441]

[156]   McHugh DJ. Production and utilization of products from commercial seaweeds. 1987.

[157]   McCandless EL, West JA, Guiry MD. Carrageenan patterns in the Gigartinaceae. Biochem Syst Ecol 1983; 11(3): 175-82.
[http://dx.doi.org/10.1016/0305-1978(83)90049-2]

[158]   De Ruiter GA, Rudolph B. Carrageenan biotechnology. Trends Food Sci Technol 1997; 8(12): 389-95.
[http://dx.doi.org/10.1016/S0924-2244(97)01091-1]

[159]   Chiovitti A, Bacic A, Craik DJ, *et al.* A pyruvated carrageenan from Australian specimens of the red alga Sarconema filiforme1Cell-wall polysaccharides from Australian red algae of the family Solieriaceae (Gigartinales, Rhodophyta). For previous instalment, see ref.[1].1. Carbohydr Res 1998; 310(1-2): 77-83.
[http://dx.doi.org/10.1016/S0008-6215(98)00170-0] [PMID: 9658565]

[160]   van de Velde F, Peppelman HA, Rollema HS, Tromp RH. On the structure of κ/ι-hybrid carrageenans. Carbohydr Res 2001; 331(3): 271-83.
[http://dx.doi.org/10.1016/S0008-6215(01)00054-4] [PMID: 11383897]

[161]   Zhou G, Sun Y, Xin H, Zhang Y, Li Z, Xu Z. *In vivo* antitumor and immunomodulation activities of different molecular weight lambda-carrageenans from *Chondrus ocellatus*. Pharmacol Res 2004; 50(1): 47-53.
        [http://dx.doi.org/10.1016/j.phrs.2003.12.002] [PMID: 15082028]

[162]   Panlasigui LN, Baello OQ, Dimatangal JM, Dumelod BD. Blood cholesterol and lipid-lowering effects of carrageenan on human volunteers. 2003; 12(2)

[163]   Cáceres PJ, Carlucci MJ, Damonte EB, Matsuhiro B, Zúñiga EA. Carrageenans from chilean samples of Stenogramme interrupta (Phyllophoraceae): structural analysis and biological activity. Phytochemistry 2000; 53(1): 81-6.
        [http://dx.doi.org/10.1016/S0031-9422(99)00461-6] [PMID: 10656412]

[164]   Carlucci MJ, Scolaro LA, Damonte EB. Inhibitory action of natural carrageenans on Herpes simplex virus infection of mouse astrocytes. Chemotherapy 1999; 45(6): 429-36.
        [http://dx.doi.org/10.1159/000007236] [PMID: 10567773]

[165]   Roberts JN, Buck CB, Thompson CD, *et al*. Genital transmission of HPV in a mouse model is potentiated by nonoxynol-9 and inhibited by carrageenan. Nat Med 2007; 13(7): 857-61.
        [http://dx.doi.org/10.1038/nm1598] [PMID: 17603495]

[166]   Morris CJ. Carrageenan-induced paw edema in the rat and mouse. Inflammation Prot 2003; pp. 115-21.
        [http://dx.doi.org/10.1385/1-59259-374-7:115]

[167]   Jeong HJ, Lee SA, Moon PD, *et al*. Alginic acid has anti-anaphylactic effects and inhibits inflammatory cytokine expression *via* suppression of nuclear factor-kappaB activation. Clin Exp Allergy 2006; 36(6): 785-94.
        [http://dx.doi.org/10.1111/j.1365-2222.2006.02508.x] [PMID: 16776680]

[168]   Kumar S, Prakash C, Chadha N, Gupta SK, Jain K, Pandey P. Effects of dietary alginic acid on growth and haemato-immunological responses of *Cirrhinus mrigala* (Hamilton, 1822) fingerlings. Turk J Fish Aquat Sci 2018; 19(5): 373-82.

[169]   Wan J, Jiang F, Xu Q, Chen D, He J. Alginic acid oligosaccharide accelerates weaned pig growth through regulating antioxidant capacity, immunity and intestinal development. RSC Advances 2016; 6(90): 87026-35.
        [http://dx.doi.org/10.1039/C6RA18135J]

[170]   Fernando IPS, Jayawardena TU, Sanjeewa KKA, Wang L, Jeon YJ, Lee WW. Anti-inflammatory potential of alginic acid from Sargassum horneri against urban aerosol-induced inflammatory responses in keratinocytes and macrophages. Ecotoxicol Environ Saf 2018; 160: 24-31.
        [http://dx.doi.org/10.1016/j.ecoenv.2018.05.024] [PMID: 29783109]

[171]   Moreno JAS, Mendes AC, Stephansen K, *et al*. Development of electrosprayed mucoadhesive chitosan microparticles. Carbohydr Polym 2018; 190: 240-7.
        [http://dx.doi.org/10.1016/j.carbpol.2018.02.062] [PMID: 29628244]

[172]   de Britto D, Assis OBG. A novel method for obtaining a quaternary salt of chitosan. Carbohydr Polym 2007; 69(2): 305-10.
        [http://dx.doi.org/10.1016/j.carbpol.2006.10.007]

[173]   Lee JY, Lee S, Choi JH, Na K. ι-Carrageenan nanocomposites for enhanced stability and oral bioavailability of curcumin. Biomater Res 2021; 25(1): 32.
        [http://dx.doi.org/10.1186/s40824-021-00236-4] [PMID: 34627398]

[174]   Sahoo P, Leong KH, Nyamathulla S, Onuki Y, Takayama K, Chung LY. Chitosan complexed carboxymethylated iota-carrageenan oral insulin particles: Stability, permeability and *in vivo* evaluation. Mater Today Commun 2019; 20: 100557.
        [http://dx.doi.org/10.1016/j.mtcomm.2019.100557]

[175]   Özbaş Z, Özkahraman B, Bayrak G, *et al*. Poly(vinyl alcohol)/(hyaluronic acid-g-kappa-carrageenan)

hydrogel as antibiotic-releasing wound dressing. Chem Zvesti 2021; 75(12): 6591-600.
[http://dx.doi.org/10.1007/s11696-021-01824-3]

[176] Madruga LYC, Popat KC, Balaban RC, Kipper MJ. Enhanced blood coagulation and antibacterial activities of carboxymethyl-kappa-carrageenan-containing nanofibers. Carbohydr Polym 2021; 273: 118541.
[http://dx.doi.org/10.1016/j.carbpol.2021.118541] [PMID: 34560953]

[177] Duan N, Li Q, Meng X, Wang Z, Wu S. Preparation and characterization of k-carrageenan/konjac glucomannan/TiO$_2$ nanocomposite film with efficient anti-fungal activity and its application in strawberry preservation. Food Chem 2021; 364: 130441.
[http://dx.doi.org/10.1016/j.foodchem.2021.130441] [PMID: 34198036]

[178] Tazhibayeva SM, Tyussyupova BB, Khamitova IK, Toktarbay Z, Musabekov KB, Daribayeva GT. Stabilization of melon cloudy juice with biopolymer agar. Eastern-European Journal of Enterprise Technologies 2020; 4(11 (106)): 31-8.
[http://dx.doi.org/10.15587/1729-4061.2020.210503]

[179] Tazhibayeva S, Tyussyupova B, Yermagambetova A, Kokanbayev A, Musabekov K. Preparation and regulation of structural-mechanical properties of biodegradable films based on starch and agar. Eastern-European Journal of Enterprise Technologies 2020; 5(6 (107)): 40-8.
[http://dx.doi.org/10.15587/1729-4061.2020.213226]

[180] Gulati K, Lal S, Kumar S, Arora S. Effect of agar and walnut (Juglans regiaL) shell fibre addition on thermal stability, water barrier, biodegradability and mechanical properties of corn starch composites. Indian Chem Engineer 2021; pp. 1-12.

[181] Jayeoye TJ, Eze FN, Singh S, Olatunde OO, Benjakul S, Rujiralai T. Synthesis of gold nanoparticles/polyaniline boronic acid/sodium alginate aqueous nanocomposite based on chemical oxidative polymerization for biological applications. Int J Biol Macromol 2021; 179: 196-205.
[http://dx.doi.org/10.1016/j.ijbiomac.2021.02.199] [PMID: 33675826]

[182] Nwabor OF, Singh S, Marlina D, Voravuthikunchai SP. Chemical characterization, release, and bioactivity of *Eucalyptus camaldulensis* polyphenols from freeze-dried sodium alginate and sodium carboxymethyl cellulose matrix. Food Quality and Safety 2020; 4(4): 203-12.
[http://dx.doi.org/10.1093/fqsafe/fyaa016]

[183] Singh S, Chunglok W, Nwabor OF, Ushir YV, Singh S, Panpipat W. Hydrophilic biopolymer matrix antibacterial peel-off facial mask functionalized with biogenic nanostructured material for cosmeceutical applications. J Polym Environ 2021.

[184] Singh S, Nwabor OF, Syukri DM, Voravuthikunchai SP. Chitosan-poly(vinyl alcohol) intelligent films fortified with anthocyanins isolated from *Clitoria ternatea* and *Carissa carandas* for monitoring beverage freshness. Int J Biol Macromol 2021; 182: 1015-25.
[http://dx.doi.org/10.1016/j.ijbiomac.2021.04.027] [PMID: 33839180]

[185] Eze FN, Jayeoye TJ, Singh S. Fabrication of intelligent pH-sensing films with antioxidant potential for monitoring shrimp freshness *via* the fortification of chitosan matrix with broken Riceberry phenolic extract. Food Chem 2022; 366: 130574.
[http://dx.doi.org/10.1016/j.foodchem.2021.130574] [PMID: 34303209]

[186] Nwabor OF, Singh S, Paosen S, Vongkamjan K, Voravuthikunchai SP. Enhancement of food shelf life with polyvinyl alcohol-chitosan nanocomposite films from bioactive Eucalyptus leaf extracts. Food Biosci 2020; 36: 100609.
[http://dx.doi.org/10.1016/j.fbio.2020.100609]

CHAPTER 4

Conformational, Morphological, and Physical Characterization of Bio-based Polymers

Abstract: Polysaccharides are the most pervasive form of pharmaceutical excipients, consisting of diverse functional properties that play a vital role in sustaining life. Moreover, polysaccharides are well-known for several benefits such as nutritional benefits, effects on immunity, and delectability with biocompatibility. Natural polysaccharides are an assembly of monosaccharides' long chain units bounded together with glycosidic linkage. In addition, polysaccharides are often quite heterogeneous, a slight alteration in the repeating unit produces distinct properties in biopolymers. Further engineered bio-based polymers produced to facilitate the regulated drug delivery system require information on structural conformation to meet the Food and Drug Administrative regulations. Furthermore, surface conformation and morphological imaging analysis are also of prime importance in the fabrication of drug delivery systems. Therefore, the amendment in the chemistry that brings about an alteration in the physicochemical property requires the use of various instrumental techniques for its characterization. In this chapter, a brief overview of compositional characterization techniques used for bio-based polymers is presented, focusing on analytical techniques that are generally applied. Moreover, the chapter promotes the application of suitable analytical techniques such as nuclear magnetic resonance spectroscopy, infrared spectroscopy, and varying chromatography in understanding the complex structure of polysaccharides. In addition, information on instruments used for surface morphological characterization of polymers is covered in this chapter.

Keywords: Bio-based polymer, Chromatography, Infrared spectroscopy, Nuclear magnetic resonance spectroscopy.

INTRODUCTION

Recent progress in biopolymers-based drug delivery technology has provided a prospect for the development of highly effective, efficient and compatible natural polymers that can regulate the release of therapeutics. Biopolymers are produced from biomaterials, where the term "bio" means biodegradable originating from a living organism. A wide variety of biomaterials resulting from biological sources such as plants and microorganisms are defined using the term bio-based polymers [1]. Whereas, excipients produced *via* the synthetic process from a biological source such as sugars, proteins, and amino acids are pronounced as modified or processed biopolymers [2]. The bio-based content of a biopolymer can be determined by calculating the number of carbon atoms that come from biomass as a raw material. Thus, biomaterials refer to materials in which carbon might emanate from non-fossil biological sources. Biopolymers are broadly classified as biodegradable and non-biodegradable polymers. Moreover, biopolymers are also classified according to occurrence and abundance as (i) polynucleotides, (ii) polypeptides/poly amino acids, and (iii) polysaccharides. Polynucleotides are long-chain polymers with 13 or more nucleotide monomers [3], while polypeptides are short polymers comprised of amino acids as monomeric units with amide bond linkage [4]. Polysaccharides are composed of sugar monomeric unit-linked *via* glycosidic linkage. Several polysaccharides have been identified and the most commonly used biopolymers are listed in Table **1**.

Table 1. Biopolymers classification [6].

Biodegradable				Non-Biodegradable
Bio-Based			**Fossil-Based**	**Bio-Based**
Plant	**Microorganism**	**Animals**		
Cellulose and derivatives	Polyhydroxy alkanoates	Chitin	Poly(alkylene dicarboxylates)	Polyethylene, polypropylene, polyvinyl chloride
Lignin	Polyhydroxy fatty acid	Chitosan	Polyglycolide	Poly(ethylene terephthalate)
Starch and derivatives	Bacterial cellulose	Hyaluronan	Poly(ε-caprolactone)	Polyurethane
Alginate	Hyaluronan	Casein	Poly(vinyl alcohol)	Polycarbonate
Lipids	Xanthan	Whey protein	Poly(ortho ester)	Poly(ether-ester)
Wheat, Corn	Curdlan	polyanhydrides	Polyanhydrides	Polyamides
Gums	Pullulan	Albumin	Polyphosphazenes	Polyester amides
Carrageenan	Silk	Keratin	-	Unsaturated polyester
Poly(lactic acid)	-	Leather	-	Epoxy

Exploring bio-based excipients, as the main component of pharmaceutical products can be a smart stratagem to mitigate cost associated issues in the development of novel drug delivery. The excellent polymeric properties of the bio-based polymers in several fields including pharmaceutical and industrial have gained a great deal of research interest. Polysaccharides are widely distributed in nature as they can be derived from plants, animals, and microorganisms. However, polysaccharides differ in their repetitive unit, and composition, which leads to a variation in physicochemical properties of biomaterials as well as fabricated pharmaceutical products. Thus, biopolymers are the promising candidate in the fabrication of a pharmaceutical product that could meet the dual requirement of sustainability and biodegradability.

Compositional characterization of bio-based polymers is considered the first step in understanding the chemistry behind the conformation of polysaccharides. Bio-based polymers are composed of monosaccharides or polysaccharides with *O*-glycosidic linkages [5]. Polysaccharides are often quite heterogeneous, hence a slight alteration in the repeating unit produces biopolymers with distinct properties. The amendment in the chemistry that brings alteration in the physicochemical property are characterized using various instrumental techniques including nuclear magnetic resonance spectroscopy, and infrared spectroscopy with various categories of chromatographic techniques. In addition, structural and chemical compositional studies of biopolymers provide information relevant for understanding compatible polymers with drug and dosage form design.

Chemistry Of Bio-Based Polymers

Advancement in the biological macromolecular structure of pharmaceutical polymers is more recent, however, less prevalent than the parallel progress in the protein, nucleic acid, and allied field [7]. Bio-based polysaccharides and gums are used in pharmaceuticals as well in textile industries due to their excellent material properties as a release retardant, suspending agent, gelling agent, wet granulator binder, fiber spinning, *etc*. The study of the molecular structure offers the most fundamental knowledge to the understanding of functional, conformational, and physiological properties of polysaccharides. However, the structural description of bio-based polymers is a perplexing task due to molecular complications. Geometric and chemical studies of biopolymers have gained significant consideration during the last decades with steady progress. The chemical characterization techniques including nuclear magnetic resonance (NMR), infrared spectroscopy (IR), chromatography, electron microscopy, x-ray diffractions, are employed for the conformational analysis of carbohydrate-derived polymers. Whereas physio-technical characterization includes moisture content, flowability, compressibility, particle size, size distribution, zeta potential, powder porosity, surface composition and morphology, contact angle and surface energy, and measurement of compressibility and compactibility.

Nuclear magnetic resonance spectroscopy is an effective tool for understanding the conformational structure, molecular dynamics, and interactional behavior of biopolymers either with therapeutics or with inactive pharmaceuticals. Combined with one-dimensional (1D) spectra, chemical shifts, and coupling coefficients in both homo and heteronuclear two dimensional (2D) NMR spectra can conjecture the association and structure of sugar residue [8]. In addition, NMR analysis allows the expert to explicate the structure of monosaccharides, oligosaccharides, glycoconjugates, and other carbohydrate derivatives either from natural or synthetic sources. Moreover, NMR has also been applied to understand the conformation, quantitative analysis, cell wall, degradation, polysaccharides mixture interaction, as well as impurities profiling in carbohydrates. Modern high-resolution NMR instruments typically of 500 MHz able to run 1D, 2D, and 3D are used for analyzing the composition and structure of carbohydrates. Compositional and structural characterization reports on bio-based polymers retrieved from the Scopus database from 1960 to 2020 demonstrated a resilient uptrend in NMR analysis of carbohydrates and derived polymers (Fig. **1**). The recent reports on the conformational chemistry of polysaccharides are presented in Table **4**.

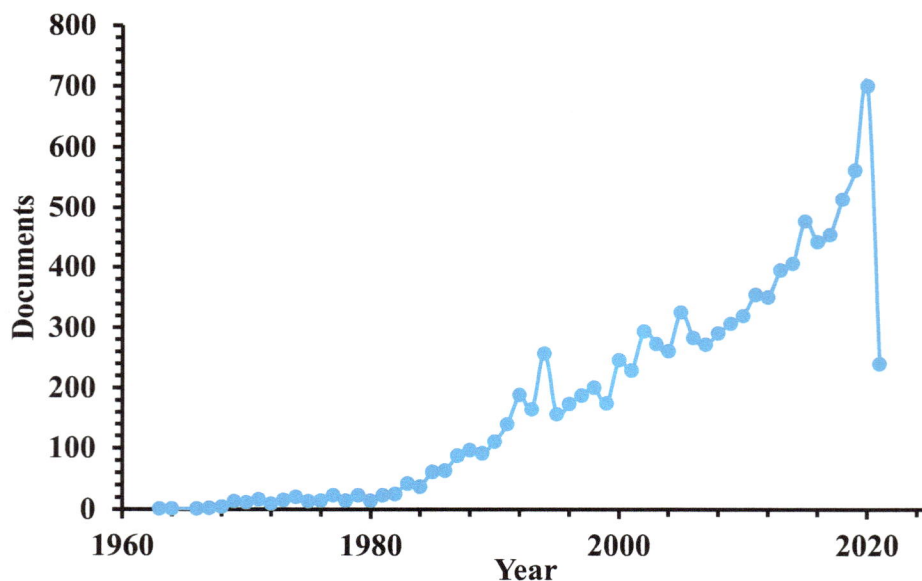

Fig (1). The annual number of documents published related to polysaccharides available online on the Scopus database from 1960 to 2020. The search includes Bullen terms such as "polysaccharides*", "Nuclear Magnetic Resonance*", "structural conformation" (Assessed on 17th June 2021).

Table 2. Fourier transforms infrared vibrational intensity and wavenumbers for the functional composition of polysaccharides [10, 11].

Vibrational Intensity/Functional Composition	Wavenumbers (Cm^{-1})
Glycoside C–O–C bending linkage	740 and 716
Sulfation on C_4 of the galactose	805

(Table 2) cont.....

Vibrational Intensity/Functional Composition	Wavenumbers (Cm^{-1})
Guluronic unit	808
Mannuronic unit	822
Sulfation on C_2 galactose	830
Anomeric CH of β-galactopyranosyl residues	890
Vibrations of the C–O–C bridge of 3,6-anhydro-L-galactose and 3,6-anhydro-D-galactose	930
Carbohydrate fingerprint	1280 – 900
Stretching vibrations of glycosidic C–O bond	1135
Asymmetric stretching of S–O	1250
Sulfate group	1370
C–H bending	1380
COO– symmetric stretching	1400
COO– asymmetric stretching	1630 – 1600
Carbonyl group of a carboxylic acid	1650
C–H stretching (symmetric or asymmetric)	3000 - 2800
O–H stretching	3500 - 3000

Table 3. Compositional characterization of recently reported bio-based polymers.

Source of polymer	Chemistry			References
	NMR	FTIR	HPLC/LC-Ms/GC-Ms	
Lentinus squarrosulus	β-pyranose, furanoses, D-galactosyl, L-fucopyranosyl, α-L-fucose, and β-D-glucose	Phospholipid, protein, β-glucans, mannose, and pyranose	-	[12]
Colocasia esculenta	Xylose, glucose, galactose, α-(1, 6)-D-Glc and β-(1, 4)-D-Glc as glucosidic bonds	Protein, carbohydrate	Arabinose, mannose, glucose, and galactose	[13, 14]
Cassia uniflora	Pyranose, xylose, galactose, β-D-glucose, β-galactopyranose, and β-D-xylopyranose,	Carbohydrate	-	[15]
Sargassum binderi	α-anomeric and α-D-linked polysaccharides	Monosaccharides	Fucose, galactose, glucose, mannose arabinose, and rhamnose	[16]
Pinus koraiensis	Pyranose, α-L-rhamnose, α-L-arabinose, β-D-mannose, β-D-glucose. β-D-galactopyranose, and hexuronic acid	Hetropolysaccharides	-	[17]
Linum usitatissimum L.	α- and β-anomeric monosaccharides	Carbohydrate	Glucose, mannose, arabinose, and xylose	[18]

(Table 3) cont.....

Source of polymer	Chemistry			References
	NMR	**FTIR**	**HPLC/LC-Ms/GC-Ms**	
Codonopsis pilosula	β-glycoside, α-arabinose, and α-glucose	Carbohydrate	Mannose, glucose, and arabinose	[19]
Pleurotus ostreatus	Heteropolysaccharides, β-glucans, α-glucans, and oligosaccharides	β-glycosidic, β-glucans, protein, and polysaccharides	-	[20]
Polygala tenuifolia	D-glactose, L-rhamnose, L-arabinose, L-mannose, and D-glucuronic acid	Pyranose	L-mannitol, L-arabitol, D-galactitol, and D-glucitol	[21]
Cucurbita moschata	Galacturonic acid, rhamnose, galactose, poly-α(1-4- -D-galacturonic acid, and α(1-2- -linked-α-L-rhamnopyranose	Rhamnose, galacturonic acid, galactose, arabinose, and xylose	-	[22]
Pseuderanthemum palatiferum	Deoxyhexose, L-rhamnose, 3-- -galacturonic acid, arabinose, and β-alactose	Mannose, arabinose, rhamnose	-	[23]
Eremurus hissaricus	β-mannose, β-D-mannose, β-D-glc, and β-D-glac	α-phyranose, and glucomannan	D-mannose, D-glucose, and D-galactose	[24]
Plantago ovata Forsk	Xylose and arabinose	Arabinoglucuronoxylan, β-xylose in pyranose, and furanose	-	[25]
Pistacia vera L.	Galacturonic acid and rhamnose	Carbohydrate	Rhamnose, glucose, galactose, mannose, xylose, arabinose, and galacturonic acid	[26]
Momordica charantia	Hexose and glucuronic acid	β-D-pyranose	Xylose, arabinose, and galactose	[27]
Opuntia macrorhiza	-	Uronic acid and overall carbohydrate	Arabinose, rhamnose, galactose and glucose	[28]
Eleocharis dulcis	Mannose, glucose, galactose, and arabinose	Monosaccharides	-	[29]
Sagittaria sagittifolia L.	Xylose, mannose, and glucose	Pyranose sugars	-	[30]
Typha domingensis	β-glycosidic anomeric proton	Polysaccharides	Arabinose, rhamnose, galactose, xylose, glucose, and mannose	[31]
Ziziphus jujuba Mil	Rhamnogalacturonan	Polysaccharides	Arabinose, galactose, glucose, mannose, and xylose	[32]
Buchanania lanzan Spreng	Arabinose, rhamnose, fructose, and mannose	Polysaccharides	Arabinose, rhamnose, fructose, and mannose	[33]
Diospyros melonoxylon Roxb	Arabinose, fructose, arabinose, mannose, rhamnose, galactose, α and β sugar residue, and α-glucose	Polysaccharides	Arabinose, fructose, mannose, and rhamnose	[34]
Manilkara zapota (Linn.)	Rhamnose, arabinose, mannose	Polysaccharides	Rhamnose, xylose, arabinose, and mannose	[35]

(Table 3) cont.....

Source of polymer	Chemistry			References
	NMR	**FTIR**	**HPLC/LC-Ms/GC-Ms**	
Ribes nigrum L	α-L-rhamnose, α-L-arabinose, α--galacturonic acid, β-D-mannose, β-D-galactose, and α-D-glucose	Pyranose	-	[36]

Table 4. Modified bio-based polymer using various techniques with pharmaceutical applications.

Source Of Polymer	Modification Techniques	Backbone Polymer	Application	Reference
Manilkara zapota	Spray drying	Hypromellose, sodium carboxymethylcellulose, polyvinyl pyrrolidone	Mucoadhesion	[2]
Buchanania lanzan	Spray drying	Mannitol, lactose, and silicon dioxide	Excipients	[40]
Tamarindus indica L	Grafting	Polyacrylamide	Flocculating material	[41, 42]
Colocasia esculenta (L.)	Grafting	Polylactide	Excipients	[43]
Plantago ovata	Grafting	Acrylic acid	Excipients	[44]
Hibiscus esculentus	Grafting/redox initiator	Acrylamide	Excipients	[45]
Coccinia indica	Grafting	Acrylamide	Excipients	[46]
Plantago psyllium	Grafting/redox initiator	Acrylonitrile	Excipients	[47]
Trigonella foenum-graecum	Grafting	Polyacrylamide	Hydrogel	[48]
Colocasia esculenta	Grafting	Poly(lactide)	Excipients	[49]
Abelmoschus esculentus	Redox initiator	Polyacrylonitrile	Excipients	[50]
Mimosa pudica	Grafting	Acrylamide	Excipients	[51]
Xanthan	Grafting	Polyacrylamide	Flooding	[52]
Chitosan	Anionic ring opening polymerization	L-Alanine-N-carboxyanhydride	Tissue engineering	[53]
Chitosan	Free radical polymerization	Acrylic acid, and β-Cyclodextrin	Controlled release of curcumin	[54]
Plantago psyllium	Redox initiator	Polyacrylamide	Flocculating agent	[55]

(Table 4) cont.....

Source Of Polymer	Modification Techniques	Backbone Polymer	Application	Reference
Plantago psyllium	Microwave assisted	Acrylic acid and acrylonitrile	Pharmaceutical beads	[56]
Trigonella foenum graecum	Redox initiator	Acrylamide	Tissue engineering	[57]
Sterculia urens	Microwave assisted	Acrylamide	Excipients	[58]
Ceratonia silique	Microwave assisted	Acrylamide	Excipients	[59]
Colocasia esculenta	Microwave assisted	Acrylamide	Excipients	[60]
Cellulose	Horseradish peroxidase	Polyacrylamide	Excipients	[61]
Xylan	Grafting	Carboxymethyl	Hydrogel	[62]

Fourier transform infrared spectroscopy (FTIR) has been extensively used as an essential technique for acquiring functional composition data of polysaccharides. FTIR vibrational spectra between wavenumber of 950 to 1200 cm^{-1} are considered as "fingerprint" region for carbohydrates [9]. Moreover, bands present in this region indicate the presence of a major functional group for specific polysaccharides for example pectin demonstrates the presence of methyl ester and amide group around 1604 and 1742 cm^{-1} [10]. In addition, FTIR spectra are also used to perceive the structural modification in polymers due to graft polymerization, co-processing of excipients, co-crystallization, *etc*. Distinctive FTIR spectra for polysaccharides are listed in Table **2**. Furthermore, recent reports on the vibrational characterization of bio-based polymers are presented in Table **3**.

Bio-based pharmaceutical polymers are used for different functional applications in drug delivery systems. However, the quality and functions of polysaccharides vary considerably due to the difference in origin, regions of production, and cultivation environments. Polysaccharides are macromolecules with a complex structure that exhibit a wide variation in molecular weight, which makes it challenging to quantify. Several chromatographic analytical techniques including thin-layer chromatography, high-performance chromatography, liquid chromatography coupled with mass spectroscopy, gas chromatography coupled with mass spectroscopy, gel chromatography, *etc*. have been used to quantify the presence of sugars (Table **2**).

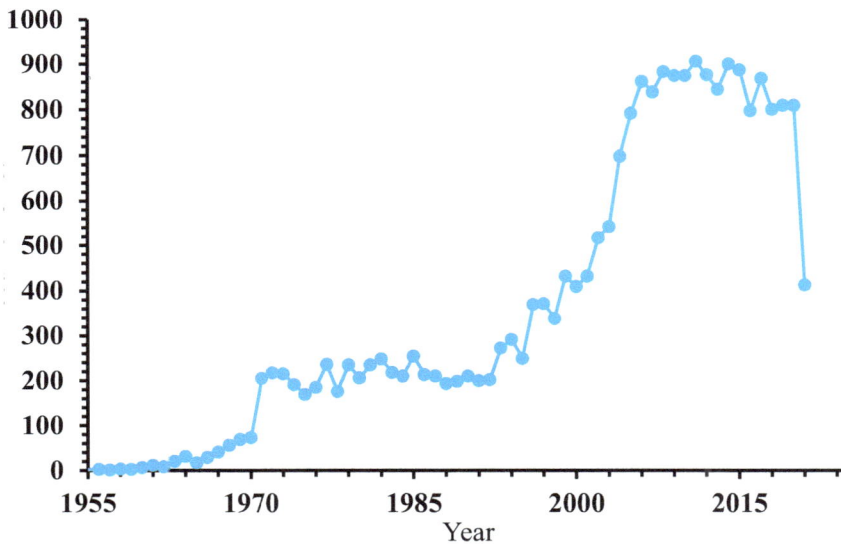

Fig (2). Total number of documents published related to graft modification of polysaccharides available online on the Scopus database from 1954 to 2020. The search includes Bullen terms such as "polysaccharides*", "grafting*", "mucilage" (Assessed on 19th June 2021).

Chemistry Of Structurally Modified Bio-Based Polymers

Polysaccharides are promising biomaterials due to their versatile property, functional modifiability, and abundance in nature [37]. Moreover, bio-based pharmaceutical excipients possess distinctive properties of biocompatibility and biodegradability. However, biopolymers indicate limitations related to loss in physicochemical properties and microbial spoilage on prolonged storage. These limitations can be overcome using various structural and functional modification methods such as co-processing, spray-drying, chemical modification, poly-electrolyte complexation, and graft polymerization. Polymer grafting is a process in which polysaccharides are covalently bonded and polymerized as side chains on the backbone of designated polymer with improvement in the polymeric properties [38]. Copolymerization modifies the symmetrical structure of the polymeric chain with modulation of both intramolecular forces and physicochemical properties such as crystallinity, solubility, glass transition temperature, permeability, elasticity, and chemical reactivity. Polymer grafting can be accomplished by various initiation processes, a few of them are ceric ion initiation, persulfate initiation, irradiation, and the commonly used hydrogen peroxide redox initiator. In addition, graft copolymerization of biopolymers is generally obtained using monomers including acrylic acid, acrylo-nitrate, acrylamide, and metha acrylate [39]. Many functional modifications have been reported in the last decades with improvement in materials properties of polysaccharides to fulfill various pharmaceutical needs as presented in Table **4**.

Fig. (**2**) demonstrates a significant surge in research related to graft structurally modified bio-based polymers as retrieved from the Scopus database from 1954 to 2020. Furthermore, Fig. (**3**) indicates the multifunctional application of grafted materials in various fields as retrieved from the Scopus database from 1954 to 2020.

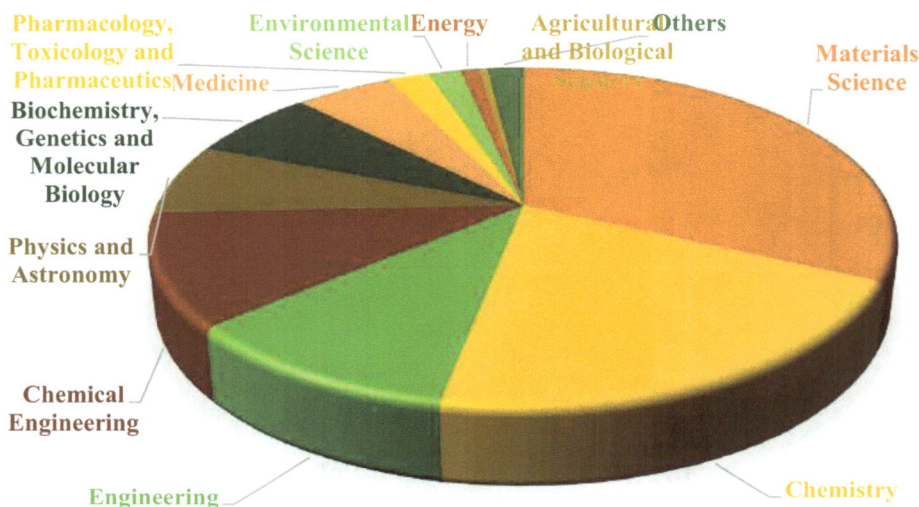

Fig (3). Application of grafted polysaccharides in the various fields available online on the Scopus database from 1954 to 2020. The search includes Bullen terms such as "polysaccharides*", "grafting*", "mucilage" (Assessed on 19th June 2021) (for interpretation of the results to color in this **Fig** legend, the reader referred to either web version of this chapter or color print).

Surface Morphology

The surface description of polymeric materials is assessed using X-ray crystallography (XRD), scan electron microscopy (SEM), and atomic force microscopy (AFM). The quantitative X-ray diffraction analysis consists of comparing the Debye-Scherrer pattern of the unknown polymeric sample, to which a known quantity of internal standard has been added. Further, the surface morphology of materials is imaged using scanning electron microscopy, which uses an electron beam to image the sample with a resolution down to the micrometer scale. Scan electron microscopy works on the principle of electron emittance from a filament and collimates into a beam in the electron source. Later the beam is focused on the sample surface by a set of lenses in the electron column under a high vacuum. This might be due to that gas molecules tend to disturb the electron beam and emit secondary and backscattered electrons. Advanced electron microscopy techniques such as transmission and field emission scanning electron microscopy (FESEM) are next-generation analytical instruments that can identify and capture microstructure images of materials at nanoscale with the quantification of specific atoms. Furthermore, the surface roughness of polymeric materials was observed *via* atomic force microscopy. Atomic force microscopy operates on the principle of surface sensing using an extremely sharp tip on a micro-machined silicon probe. The surface image of material is captured using a tip by raster scanning across the surface line by line. The material surface roughness plays a pivotal role in understanding the physics of tableting technology. The physics of tablets deliver a relationship between upper and lower punch forces with the ejection forces during the process of tablet compression either dry or wet granulation. Additionally, the surface roughness of materials helps to understand the bonding phenomena among the particles of polymers themselves and drug molecules.

Physio-Technical Characterization

The physical and technical characterizations of a biopolymer are necessary to understand its suitability for the development of dosage form. The physico-technical characterization includes moisture content, flowability, compressibility, particle size, size distribution, zeta potential, powder porosity, surface composition, morphology, contact angle, surface energy, measurement of compressibility, compactibility, *etc*.

The moisture content of polymers is estimated manually as well as using a sophisticated moisture analyzer. The automatic mode of moisture analyzer investigates the difference in sample weight changes by less than 1 mg/min. The flowability of biopolymers is quantified by pouring it through a vibrating metal funnel onto a platform until a stable and height-fixed heap forms. Whereas, the angle of repose is measured as an angle made by the inclined plane of the heap with the horizontal. Moreover, the compressibility of polymers is measured *via* bulk density by pouring polymer samples through a vibrating metal funnel into a measuring cylinder, and the volume of the cylinder containing powder/granules is recorded. In addition, tap density is measured by recording the appropriate sample height in the cylinder after tapping it. Furthermore, the porosity of polymeric materials is computed using results of material bulk and true density.

The particle size measurement technically determines the size range, average, or mean size of the particles in powder or granular samples. Traditionally, particle size is determined using sieves and sedimentation analysis, however, the modern analytical tools based on different technologies such as electrical sensing zone, chromatography techniques, high definition image processing, analysis of Brownian motion, gravitational settling, and light scattering of particles are commonly used by pharmaceutical researchers and manufactures. Recently, particle size analysis based on light scattering demonstrated widespread application in various fields, including pharmaceutical, food, cosmetic, and polymer production. The particle size measurement *via* scattering of light works on the principle of sample illumination by a laser beam and the fluctuation of the scattered light detected at a known scattering angle θ using a fast photon detector. Zeta potential is measured to identify the type of surface charge present on the materials. The quantification of surface charge provides information on the suitability of polymer in the fabrication of various conventional or novel drug delivery systems.

CONCLUSIONS AND FUTURE PROSPECTS

Polysaccharides are complex biological macromolecules that demonstrate several potential biological and pharmacological activities. Structural and morphological analysis of such a biopolymer plays a vital role in the elucidation of compatibility and structure-activity relationships. Nuclear Magnetic Resonance and Infrared Spectroscopy with various chromatography methods are extensively used analytical techniques in the structural and conformational exploration of biopolymers. In addition, engineered biopolymers produced to facilitate the regulated drug delivery system also demand data on structural conformation to meet the Food and Drug Administrative regulatory norms. Moreover, conformational characterization of bio-based polymers offers a foundation that helps researchers to understand the chemistry of polysaccharides suitable for drug delivery systems. In addition, morphological characterization using X-ray diffraction, SEM, FESEM, TEM, and AFM are of potential importance in the interpretation of polymeric materials. Although, analytical techniques have advanced in the quantification of polysaccharides conformational structure, however, structural conformation relationship cannot be identified at a molecular level. Thus, future development in conformational analysis of engineered bio-based polymers requires improvement in NMR chemical shift assignments that can be resolved by calculating the potential mean force surface of oligosaccharides using molecular dynamics simulations [63 - 65]. Moreover, improvement in methylated product NMR and IR spectra records could also support the conformational characterization of polysaccharides by developing a distinct "fingerprint" region [65]. On the other hand, complex polysaccharides with distinct molecular weight can be characterized and differentiated using modern chromatographic techniques.

REFERENCES

[1]   Ige OO, Umoru LE, Aribo S. Natural Products: A Minefield of Biomaterials. ISRN Materials Science 2012; 2012: 1-20.
[http://dx.doi.org/10.5402/2012/983062]

[2]   Singh S, Nwabor OF, Ontong JC, Kaewnopparat N, Voravuthikunchai SP. Characterization of a novel, co-processed bio-based polymer, and its effect on mucoadhesive strength. Int J Biol Macromol 2020;

145: 865-75.
[http://dx.doi.org/10.1016/j.ijbiomac.2019.11.198] [PMID: 31783076]

[3]   Kumar V, Chibuzo EN, Garza-Reyes JA, Kumari A, Rocha-Lona L, Lopez-Torres GC. The impact of supply chain integration on performance: evidence from the UK food sector. Procedia Manuf 2017; 11: 814-21.
[http://dx.doi.org/10.1016/j.promfg.2017.07.183]

[4]   Numata K. Poly(amino acid)s/polypeptides as potential functional and structural materials. Polym J 2015; 47(8): 537-45.
[http://dx.doi.org/10.1038/pj.2015.35]

[5]   Song EH, Shang J, Ratner DM. Polysaccharides. In: K Matyjaszewski, M Möller, Eds. Polymer Science: a comprehensive reference. Amsterdam Elsevier 2012; pp. 137-55.

[6]   Niaounakis M. Definitions of terms and types of biopolymers. In: Niaounakis M, Ed. Biopolymers: applications and trends. Oxford William Andrew Publishing 2015; pp. 1-90.
[http://dx.doi.org/10 1016/B978-0-323-35399-1.00001-6]

[7]   Liechty WB, Kryscio DR, Slaughter BV, Peppas NA. Polymers for drug delivery systems. Annu Rev Chem Biomol Eng 2010; 1(1): 149-73.
[http://dx.doi.org/10 1146/annurev-chembioeng-073009-100847] [PMID: 22432577]

[8]   Benito C. Nuclear magnetic resonance studies of polysaccharide structure and interactions. In: Atkins EDT, Ed. Polysaccharides topics in molecular and structural biology London The MACMillan Press LTD. 1985; pp. 1-40.

[9]   Černá M, Barros AS, Nunes A, *et al.* Use of FT-IR spectroscopy as a tool for the analysis of polysaccharide food additives. Carbohydr Polym 2003; 51(4): 383-9.
[http://dx.doi.org/10 1016/S0144-8617(02)00259-X]

[10]   Guo QAL, Cui SW. Fourier transform infrared spectroscopy (FTIR) for carbohydrate analysis. In: Navard P, Ed. Methodology for structural analysis of polysaccharides New York City. Cham: Springer 2018; pp. 69-71.
[http://dx.doi.org/10 1007/978-3-319-96370-9_9]

[11]   Fernando IPS, Sanjeewa KKA, Samarakoon KW, *et al.* FTIR characterization and antioxidant activity of water soluble crude polysaccharides of Sri Lankan marine algae. Algae 2017; 32(1): 75-86.
[http://dx.doi.org/10.4490/algae.2017.32.12.1]

[12]   Ayimbila F, Siriworg S, Keawsompong S. Structural characteristics and bioactive properties of water-soluble polysaccharide from *Lentinus squarrosulus*. Bioactive Carbohydrates and Dietary Fibre 2021; 26: 100266.
[http://dx.doi.org/10.1016/j.bcdf.2021.100266]

[13]   Anwar M, McConnell M, Bekhit AED. New freeze-thaw method for improved extraction of water-soluble non-starch polysaccharide from taro (*Colocasia esculenta*): Optimization and comprehensive characterization of physico-chemical and structural properties. Food Chem 2021; 349: 129210.
[http://dx.doi.org/10.1016/j.foodchem.2021.129210] [PMID: 33582541]

[14]   Li H, Dong Z, Liu X, Chen H, Lai F, Zhang M. Structure characterization of two novel polysaccharides from *Colocasia esculenta* (taro) and a comparative study of their immunomodulatory activities. J Funct Foods 2018; 42: 47-57.
[http://dx.doi.org/10.1016/j.jff.2017.12.067]

[15]   Deore UV, Mahajan HS. Isolation and structural characterization of mucilaginous polysaccharides obtained from the seeds of *Cassia uniflora* for industrial application. Food Chem 2021; 351: 129262.
[http://dx.doi.org/10.1016/j.foodchem.2021.129262] [PMID: 33626466]

[16]   Je JG, Lee HG, Fernando KHN, Jeon YJ, Ryu B. Purification and structural characterization of sulfated polysaccharides derived from brown algae, *Sargassum binderi*: inhibitory mechanism of iNOS and COX-2 pathway interaction. Antioxidants 2021; 10(6): 822.

[http://dx.doi.org/10.3390/antiox10060822] [PMID: 34063885]

[17] Zhang H, Zou P, Zhao H, Qiu J, Regenstein JM, Yang X. Isolation, purification, structure and antioxidant activity of polysaccharide from pinecones of *Pinus koraiensis*. Carbohydr Polym 2021; 251: 117078.
[http://dx.doi.org/10.1016/j.carbpol.2020.117078] [PMID: 33142621]

[18] Trabelsi I, Ben Slima S, Ktari N, Bouaziz M, Ben Salah R. Structure analysis and antioxidant activity of a novel polysaccharide from katan seeds. BioMed Res Int 2021; 2021: 1-13.
[http://dx.doi.org/10.1155/2021/6349019] [PMID: 33511204]

[19] Yuan S, Xu CY, Xia J, Feng YN, Zhang XF, Yan YY. Extraction of polysaccharides from *Codonopsis pilosula* by fermentation with response surface methodology. Food Sci Nutr 2020; 8(12): 6660-9.
[http://dx.doi.org/10.1002/fsn3.1958] [PMID: 33312549]

[20] Barbosa JRS, S Freitas MM, Oliveira LCS, *et al.* Obtaining extracts rich in antioxidant polysaccharides from the edible mushroom Pleurotus ostreatus using binary system with hot water and supercritical CO_2. Food Chem 2020; 330: 127173.
[http://dx.doi.org/10.1016/j.foodchem.2020.127173] [PMID: 32569930]

[21] Li J, Zhong J, Chen H, Yu Q, Yan C. Structural characterization and anti-neuroinflammatory activity of a heteropolysaccharide isolated from the rhizomes of *Polygala tenuifolia*. Ind Crops Prod 2020; 155: 112792.
[http://dx.doi.org/10.1016/j.indcrop.2020.112792]

[22] Thuy TTT, Bien DT, Thu QTM, Xuan DTT, Van Nguyen B, Van Quang N, *et al.* Extraction and structural determination of pectin from pumkin *Cucurbita moschata*. Vietnam J Chem 2020; 58(5): 592-6.

[23] Ho TC, Kiddane AT, Sivagnanam SP, *et al.* Green extraction of polyphenolic-polysaccharide conjugates from *Pseuderanthemum palatiferum* (Nees) Radlk.: Chemical profile and anticoagulant activity. Int J Biol Macromol 2020; 157: 484-93.
[http://dx.doi.org/10.1016/j.ijbiomac.2020.04.113] [PMID: 32325075]

[24] Muhidinov ZK, Bobokalonov JT, Ismoilov IB, *et al.* Characterization of two types of polysaccharides from *Eremurus hissaricus* roots growing in Tajikistan. Food Hydrocoll 2020; 105: 105768.
[http://dx.doi.org/10.1016/j.foodhyd.2020.105768]

[25] Ren Y, Yakubov GE, Linter BR, MacNaughtan W, Foster TJ. Temperature fractionation, physicochemical and rheological analysis of psyllium seed husk heteroxylan. Food Hydrocoll 2020; 104: 105737.
[http://dx.doi.org/10.1016/j.foodhyd.2020.105737]

[26] Hamed M, Bougatef H, Karoud W, *et al.* Polysaccharides extracted from pistachio external hull: Characterization, antioxidant activity and potential application on meat as preservative. Ind Crops Prod 2020; 148: 112315.
[http://dx.doi.org/10.1016/j.indcrop.2020.112315]

[27] Yang X, Chen F, Huang G. Extraction and analysis of polysaccharide from Momordica charantia. Ind Crops Prod 2020; 153: 112588.
[http://dx.doi.org/10.1016/j.indcrop.2020.112588]

[28] Amamou S, Lazreg H, Hafsa J, *et al.* Effect of extraction condition on the antioxidant, antiglycation and α-amylase inhibitory activities of *Opuntia macrorhiza* fruit peels polysaccharides. Lebensm Wiss Technol 2020; 127: 109411.
[http://dx.doi.org/10.1016/j.lwt.2020.109411]

[29] Zeng F, Chen W, He P, *et al.* Structural characterization of polysaccharides with potential antioxidant and immunomodulatory activities from Chinese water chestnut peels. Carbohydr Polym 2020; 246: 116551.
[http://dx.doi.org/10.1016/j.carbpol.2020.116551] [PMID: 32747236]

[30]   Gu J, Zhang H, Wen C, *et al.* Purification, characterization, antioxidant and immunological activity of polysaccharide from *Sagittaria sagittifolia* L. Food Res Int 2020; 136: 109345.
[http://dx.doi.org/10.1016/j.foodres.2020.109345] [PMID: 32846537]

[31]   Sorourian R, Khajehrahimi AE, Tadayoni M, Azizi MH, Hojjati M. Ultrasound-assisted extraction of polysaccharides from *Typha domingensis*: Structural characterization and functional properties. Int J Biol Macromol 2020; 160: 758-68.
[http://dx.doi.org/10.1016/j.ijbiomac.2020.05.226] [PMID: 32485259]

[32]   Ji X, Hou C, Yan Y, Shi M, Liu Y. Comparison of structural characterization and antioxidant activity of polysaccharides from jujube (*Ziziphus jujuba* Mill) fruit. Int J Biol Macromol 2020; 149: 1008-18.
[http://dx.doi.org/10.1016/j.ijbiomac.2020.02.018] [PMID: 32032709]

[33]   Singh S, Bothara S. Morphological, physico-chemical and structural characterization of mucilage isolated from the seeds of *Buchanania lanzan* Spreng. Int J Health Allied Sci 2014; 3(1): 33.
[http://dx.doi.org/10.4103/2278-344X.130609]

[34]   Singh S, Bothara SB. Physico-chemical and structural characterization of mucilage isolated from seeds of *Diospyros melonoxylon* Roxb. Braz J Pharm Sci 2014; 50(4): 713-25.
[http://dx.doi.org/10.1590/S1984-82502014000400006]

[35]   Singh S, Bothara SB. *Manilkara zapota* (Linn.) Seeds: A potential source of natural gum. ISRN Pharm 2014; 2014: 1-10.
[http://dx.doi.org/10.1155/2014/647174] [PMID: 24729907]

[36]   Yang Y, Lei Z, Zhao M, Wu C, Wang L, Xu Y. Microwave-assisted extraction of an acidic polysaccharide from *Ribes nigrum* L.: Structural characteristics and biological activities. Ind Crops Prod 2020; 147: 112249.
[http://dx.doi.org/10.1016/j.indcrop.2020.112249]

[37]   Song HQ, Fan Y, Hu Y, Cheng G, Xu FJ. Polysaccharide peptide conjugates: a versatile material platform for biomedical applications. Adv Funct Mater 2021; 31(6): 2005978.
[http://dx.doi.org/10.1002/adfm.202005978]

[38]   Choudhary S, Sharma K, Sharma V, Kumar V. Grafting Polymers. In: Gutierrez TJ, Ed. Reactive and functional polymers. Springer Cham, Springer Nature Switzerland AG 2020; pp. 199-243.
[http://dx.doi.org/10.1007/978-3-030-45135-6_8]

[39]   Gürdağ G, Sarmad S. 3r: synthesis, properties, and application. In: Susheel K, Sabaa MW, Eds. Polysaccharide based graft copolymers Springer. Berlin: Heiderberg 2013; pp. 15-57.
[http://dx.doi.org/10.1007/978-3-642-36566-9_2]

[40]   Singh S, Nwabor OF, Ontong JC, Voravuthikunchai SP. Characterization and assessment of compression and compactibility of novel spray-dried, co-processed bio-based polymer. J Drug Deliv Sci Technol 2020; 56: 101526.
[http://dx.doi.org/10.1016/j.jddst.2020.101526]

[41]   Nandi G, Changder A, Ghosh LK. Graft-copolymer of polyacrylamide-tamarind seed gum: Synthesis, characterization and evaluation of flocculating potential in peroral paracetamol suspension. Carbohydr Polym 2019; 215: 213-25.
[http://dx.doi.org/10.1016/j.carbpol.2019.03.088] [PMID: 30981348]

[42]   Mishra A, Malhotra AV. Graft copolymers of xyloglucan and methyl methacrylate. Carbohydr Polym 2012; 87(3): 1899-904.
[http://dx.doi.org/10.1016/j.carbpol.2011.09.068]

[43]   Mijinyawa AH, Durga G, Mishra A. Isolation, characterization, and microwave assisted surface modification of *Colocasia esculenta* (L.) Schott mucilage by grafting polylactide. Int J Biol Macromol 2018; 119: 1090-7.
[http://dx.doi.org/10.1016/j.ijbiomac.2018.08.045] [PMID: 30099042]

[44]   Rao MRP, Babrekar L, Kharpude VS, Chaudhari J. Synthesis and characterization of psyllium seed

mucilage grafted with N, N -methylene bisacrylamide. Int J Biol Macromol 2017; 103: 338-46.
[http://dx.doi.org/10.1016/j.ijbiomac.2017.05.031] [PMID: 28512054]

[45]   Mishra A, Clark JH, Pal S. Modification of Okra mucilage with acrylamide: Synthesis, characterization and swelling behavior. Carbohydr Polym 2008; 72(4): 608-15.
[http://dx.doi.org/10.1016/j.carbpol.2007.10.009]

[46]   Mishra A, Bajpai M. Synthesis and characterization of polyacrylamide grafted copolymers of Kundoor mucilage. J Appl Polym Sci 2005; 98(3): 1186-91.
[http://dx.doi.org/10.1002/app.22173]

[47]   Mishra A, Srinivasan R, Gupta R. P. psyllium-g-polyacrylonitrile: synthesis and characterization. Colloid Polym Sci 2003; 281(2): 187-9.
[http://dx.doi.org/10.1007/s00396-002-0777-x]

[48]   Hussain HR, Bashir S, Mahmood A, *et al.* Fenugreek seed mucilage grafted poly methacrylate pH-responsive hydrogel: A promising tool to enhance the oral bioavailability of methotrexate. Int J Biol Macromol 2022; 202: 332-44.
[http://dx.doi.org/10.1016/j.ijbiomac.2022.01.064] [PMID: 35041883]

[49]   Mijinyawa AH, Durga G, Mishra A. Evaluation of thermal degradation and melt crystallization behavior of taro mucilage and its graft copolymer with poly(lactide). SN Applied Sciences 2019; 1(11): 1486.
[http://dx.doi.org/10.1007/s42452-019-1490-4]

[50]   Mishra A, Pal S. Polyacrylonitrile-grafted Okra mucilage: A renewable reservoir to polymeric materials. Carbohydr Polym 2007; 68(1): 95-100.
[http://dx.doi.org/10.1016/j.carbpol.2006.07.014]

[51]   Ahuja M, Kumar S, Yadav M. Microwave-assisted synthesis and characterization of polyacrylamide grafted co-polymers of Mimosa mucilage. Polym Bull 2011; 66(9): 1163-75.
[http://dx.doi.org/10.1007/s00289-010-0341-7]

[52]   Chami S, Joly N, Bocchetta P, Martin P, Aliouche D. Polyacrylamide grafted xanthan: microwave-assisted synthesis and rheological behavior for polymer flooding. Polymers (Basel) 2021; 13(9): 1484.
[http://dx.doi.org/10.3390/polym13091484] [PMID: 34063011]

[53]   Park GH, Kang MS, Knowles JC, Gong MS. Synthesis, characterization, and biocompatible properties of alanine-grafted chitosan copolymers. J Biomater Appl 2016; 30(9): 1350-61.
[http://dx.doi.org/10.1177/0885328215626892] [PMID: 26767393]

[54]   Anirudhan TS, Divya PL, Nima J. Synthesis and characterization of novel drug delivery system using modified chitosan based hydrogel grafted with cyclodextrin. Chem Eng J 2016; 284: 1259-69.
[http://dx.doi.org/10.1016/j.cej.2015.09.057]

[55]   Mishra A, Srinivasan R, Bajpai M, Dubey R. Use of polyacrylamide-grafted Plantago psyllium mucilage as a flocculant for treatment of textile wastewater. Colloid Polym Sci 2004; 282(7): 722-7.
[http://dx.doi.org/10.1007/s00396-003-1003-1]

[56]   Kumar D, Pandey J, Kumar P. Microwave assisted synthesis of binary grafted psyllium and its utility in anticancer formulation. Carbohydr Polym 2018; 179: 408-14.
[http://dx.doi.org/10.1016/j.carbpol.2017.09.093] [PMID: 29111068]

[57]   Bal T, Swain S. Microwave assisted synthesis of polyacrylamide grafted polymeric blend of fenugreek seed mucilage-Polyvinyl alcohol (FSM-PVA-g-PAM) and its characterizations as tissue engineered scaffold and as a drug delivery device. Daru 2020; 28(1): 33-44.
[http://dx.doi.org/10.1007/s40199-019-00237-8] [PMID: 30712231]

[58]   Kaur L, Gupta G. Gum karaya-g-poly (acrylamide): microwave assisted synthesis, optimisation and characterisation. Int J Appl Pharm 2020; pp. 143-52.

[59]   Giri TK, Pure S, Tripathi DK. Synthesis of graft copolymers of acrylamide for locust bean gum using microwave energy: swelling behavior, flocculation characteristics and acute toxicity study. Polímeros

2015; 25(2): 168-74.
[http://dx.doi.org/10 1590/0104-1428.1717]

[60]   Loveleenpreet Kau⁻ GGD. Gum colocasia-g-polyacrylamide: Microwave-assisted synthesis and characterization. International Journal of Research in Pharmaceutical Sciences 2020; 11(4): 6698-706.
[http://dx.doi.org/10 26452/ijrps.v11i4.3595]

[61]   Wang R, Rong L, Ni S, *et al.* Enzymatic graft polymerization from cellulose acetoacetate: a versatile strategy for cellulose functionalization. Cellulose 2021; 28(2): 691-701.
[http://dx.doi.org/10.1007/s10570-020-03577-w]

[62]   Li N, Sun D, Su Z, *et al.* Rapid fabrication of xylan-based hydrogel by graft polymerization *via* a dynamic lignin-Fe^{3+} plant catechol system. Carbohydr Polym 2021; 269: 118306.
[http://dx.doi.org/10.1016/j.carbpol.2021.118306] [PMID: 34294323]

[63]   Yang M, Angles c'Ortoli T, Säwén E, Jana M, Widmalm G, MacKerell AD. Delineating the conformational flexibility of trisaccharides from NMR spectroscopy experiments and computer simulations. Phys Chem Chem Phys 2016; 18(28): 18776-94.
[http://dx.doi.org/10.1039/C6CP02970A] [PMID: 27346493]

[64]   Gray CJ, Migas LG, Barran PE, *et al.* Advancing Solutions to the Carbohydrate Sequencing Challenge. J Am Chem Soc 2019; 141(37): 14463-79.
[http://dx.doi.org/10.1021/jacs.9b06406] [PMID: 31403778]

[65]   Yao HYY, Wang JQ, Yin JY, Nie SP, Xie MY. A review of NMR analysis in polysaccharide structure and conformation: Progress, challenge and perspective. Food Res Int 2021; 143: 110290.
[http://dx.doi.org/10.1016/j.foodres.2021.110290] [PMID: 33992390]

Thermo-Mechanical Properties of Bio-Based Polymers

Abstract: Bio-based polymers offer a broad range of applications in pharmaceutical engineering. However, their assortment gets constrained owing to variations in structural conformation, which affects the thermomechanical properties during complex formulation. The thermomechanical property of pharmaceutically inactive ingredients provides insight into the thermal expansion, glass transitions temperature, softening point, compositional, and phase changes of biomaterials with different geometries on the application of constant force as a function of temperature. In addition, thermomechanical properties provide fundamental information on network chemical structure, crosslink density, rubbery modulus, failure strain, and toughness. Moreover, the structural composition of polysaccharides also affects the composite's mechanical properties. Hence, analysis of thermomechanical properties provides valuable information that is applicable in different sectors including aviation, quasi-static loading, electroplating technology, micro-electric, construction, cosmetics, food packaging, and pharmaceutical products. This compilation highlights the basics of thermal and mechanical experiments on bio-based polymers with different fabrication for both technical and pharmaceutical formulations.

Keywords: Bio-based polymers, Differential scanning calorimetric, Differential thermal analysis, Thermogravimetric analysis, Thermomechanical.

INTRODUCTION

Natural resources, both renewable and non-renewable, significantly contribute to monetary benefits, *via* utilization in various manufacturing processes. The World Trade Organization, in 2010 recognized that natural resources are stocks of materials, that are both scarce and economically useful in production or consumption, either in their raw state or after a minimal expense of processing [1]. Geographic surface coverage has broadly classified natural resources into two types – *Point Resources* and *Diffused Resources*. The point resources include a narrow economic base such as oils and minerals, whereas diffused resources include agriculture and forest [2]. Moreover, diffused natural resources are a class of renewable natural sources that significantly contribute to the development of herbal therapeutics. Considering herbal therapeutics, active and inactive pharmac-

Sudarshan Singh & Warangkana Chunglok

eutical ingredients have gained importance due to their cytocompatibility and biodegradability.

Polysaccharides are pharmaceutical natural polymers with distinctive configurations and properties favorable in several biomedical applications. Polysaccharides are isolated from the plant, animal, microbial, and marine sources. Polysaccharides are carbohydrates that are comprised of monosaccharides attached with O-glycosidic linkages and are essential biomolecules required for the growth of living organisms [3]. Moreover, polysaccharides can be homogenous, containing only one kind of sugars, whereas heterogeneous polysaccharides contain 8-10 types of sugar moiety [4]. Among them, plant polysaccharides have gained enormous attention as bio-based polymers with a wide range of applications in pharmaceuticals, prosthetics, imaging analysis, and food industries [1] as presented in Fig. (**1** and **2**). Recent reports indicated that more than 300 different kinds of bio-based polymers have been identified for drug delivery applications [5]. Moreover, several biomaterials have been chemically modified to regulate the release of active ingredients as well as to support the requirement of formulations.

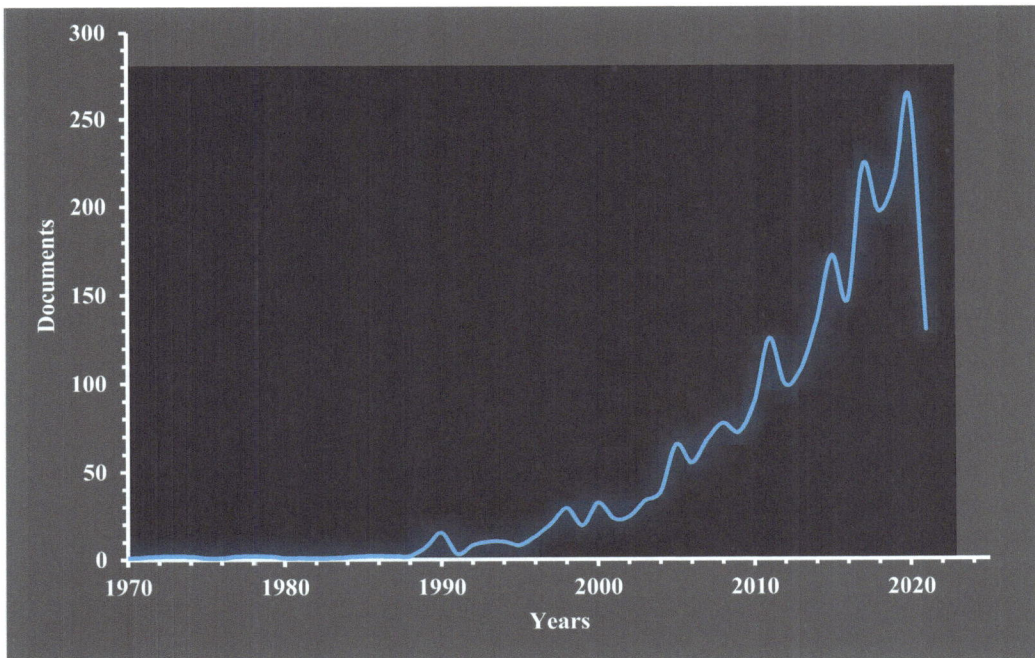

Fig. (1). Publication of documents from 1970 to 2020, data retrieved from Scopus database on 07th of June 2021 using keywords "biodegradable*" "polysaccharides*".

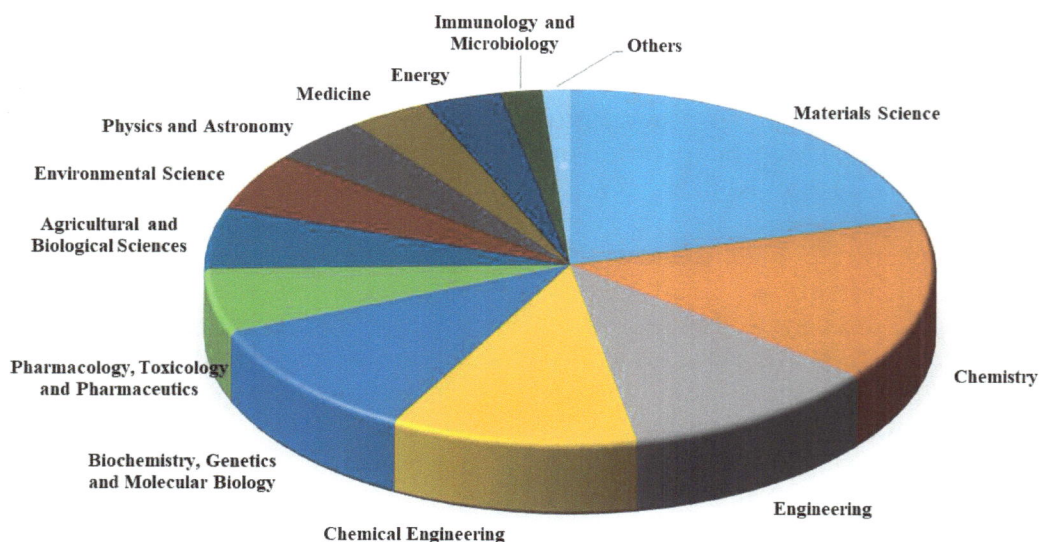

Fig. (2). Publication of documents on various subjects from 1970 to 2020, data retrieved from Scopus database on 07th of June 2021 using keywords "biodegradable*" "polysaccharides*"(for interpretation of the results to color in this Fig. legend, the reader referred to either web version of this chapter or color print).

The pharmaceutical formulation consists of both active and inactive constituents where the individual may experience phase transitions during formulation and storage in the varied range of temperature and pressure. The molecular changes exerted during the phase transitions promote variation in the thermomechanical properties. Therefore, the phase transitions analysis of pharmaceutical formulation is of major importance in maintaining the integrity of the product during shelf life. The phase transitional thermal analysis technologies include thermogravimetric analysis (TGA), differential scanning calorimetry (DSC), and differential thermal analysis (DTA).

Thermal Analysis

Thermal analysis of pharmaceutical excipients includes a series of temperature dependency measurements aimed to understand the physicochemical property of the substance. Several techniques can be employed for the measurement of thermo-analytical properties as listed in Table **1**, however only the most popular techniques used for characterization of excipients have been elaborated in this chapter.

Table 1. List of thermo-analytical methods used to analyze thermal property of pharmaceutical materials [6].

Methods	Application
Differential thermal analysis (DTA)	Temperature difference (ΔT)
Differential scanning calorimetry (DSC)	Change in enthalpy
Thermogravimetric analysis (TGA/Tg)	Mass loss
Micro/Nano-thermal analysis (μ/n-TA)	Penetration, (ΔT)
Dynamic mechanical analysis (DMA)	Deformation
Dielectric thermal analysis (DEA)	Deformation
Evolved gas analysis (EGA)	Gaseous decomposition
Thermo-optical analysis (TOA)	Optical properties

A differential thermal instrument measures the temperature difference among sample and reference materials with allusion to temperature in a programmed atmosphere. The term differential is used since alterations in the specimens are measured with reference to standard materials for a fixed amount of heat inputs (Fig. **3**). Differential thermal analysis and DSC are similar in many instances, and comparable thermal events can be observed in both cases [7]. However, DTA differs from DSC in ways including the determination of temperature difference in DTA, whereas DSC measures enthalpy changes. Although, DTA requires a lower quantity of samples then also it is four times higher than DSC. Moreover, DTA utilizes an open alumina crucible, whereas DSC is conducted in a closed crucible. Since DTA works on dynamic performances, the experimental curve is affected by several factors depending on instruments' parameter and sample characteristics. Instrumental factor includes furnace atmosphere, heating rate, and location of thermocouples in the sample chamber, whereas the degree of crystallinity, presence of diluents, thermal conductivity, and heat capacity are related to sample characteristics [8].

Differential scanning calorimetry came into existence in 1963, under the manufacturing excellence of Perkin-Elmer [9]. According to ASTM-E437 standard, DSC works on the principle of heat flow difference between sample and reference, measured as a function of temperature, under controlled conditions as presented in Fig. (**4**). Differential scanning calorimeter measures both qualitative and quantitative heat content information either in an endothermic or exothermic form in J/mol due to physicochemical changes of tested materials. Thermal analysis of biomaterials is affected basically by two factors; (I) instrumental factors and (II) those related to sample characteristics. Instrumental factors are the geometry of the sample holder, furnace-heating rate, the sensitivity of recording system with speed, and furnace atmosphere. While the weight of the sample, particle size, the solubility of evolved gases in the sample, and heat of reaction with thermal conductivity are related to sample physiognomies [8].

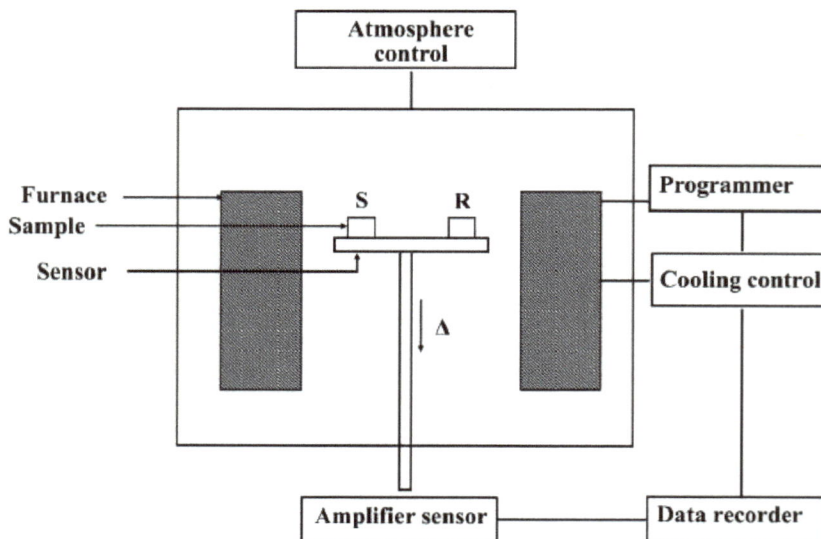

Fig. (3). Schematic of differential thermal analysis.

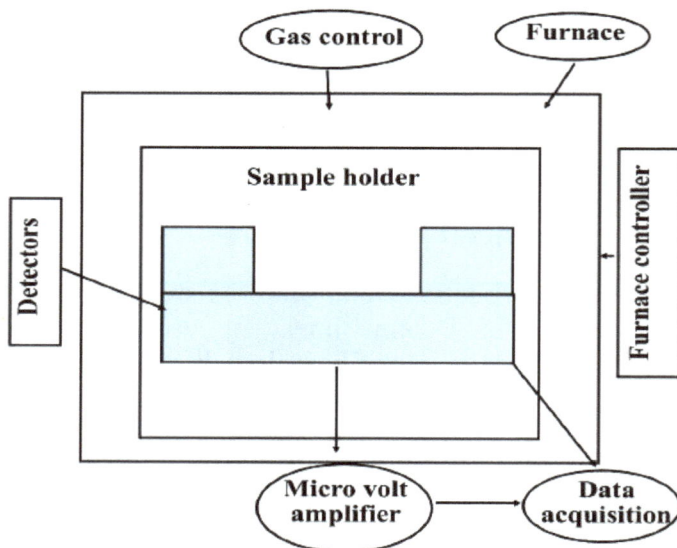

Fig. (4). Schematic image of differential scanning calorimetric.

The thermogravimetric analysis measures changes in mass due to changes in the physicochemical process upon heating a sample under controlled temperature program and thermosphere [10]. Currently, several TGA instruments with variation in the design are available including high precise balance to which sample holder pan is attached (Fig. **5**). Moreover, a few commercial instruments available allow simultaneous application of TGA-DTA (STGA) and TGA-DSC (STDSC) for a single sample with extra accuracy as presented in Fig. (**6**). Thermogravimetric instruments primarily work at ambient temperature of 1000 °C, however, some excipients such as *silica* that indicates mass loss above ambient temperature require maneuver till 1300 °C [11]. TGA works under a purge of an inert gas such as nitrogen, argon, or helium; oxidizing air such as oxygen, or reducing gas such as forming gas (8-10% hydrogen in nitrogen). Subsequently, the moisture content of purge varies from dry to saturate depending on materials requirements. In addition, the next-generation TGA works on integrated isothermal or dynamic heating and cooling cycles, programed to hold the process at a specific temperature once a change in mass is detected. Bio-based polymers generally exhibit mass loss, as volatile degradation components such as absorbed moisture, residual solvents, or low molecular mass additives evaporate between 25 – 800 °C with charring and the formation of ash. The variation in the overall thermo-analytical profile of pharmaceutical materials is affected by several factors during TGA analysis as listed in Table **2**. Furthermore, a list of recent bio-based polymers tested for mass loss using TGA is presented in Table **3**.

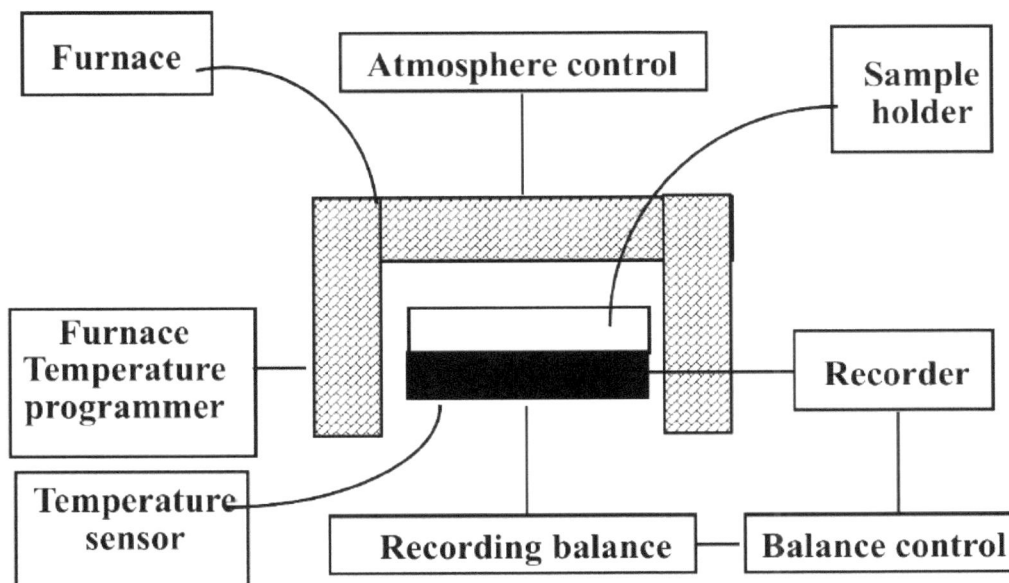

Fig. (5). Block diagram of thermogravimetric analysis.

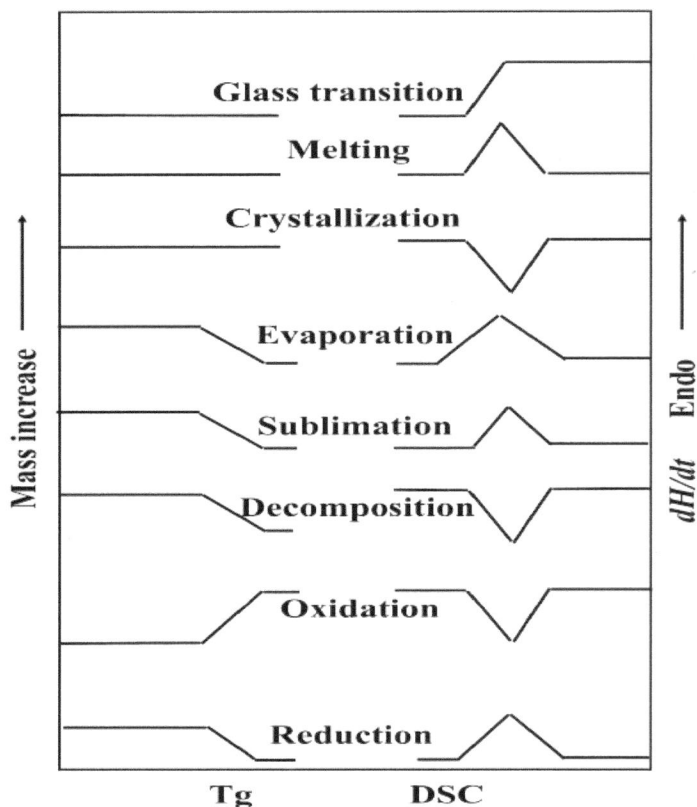

Fig. (6). Simultaneous thermogravimetric and differential scanning thermal analysis (6).

Table 2. Factor affecting processing of thermogravimetric analysis of pharmaceutical substances.

Mass	Temperature
Thermal expansion	Heating rate
Turbulence due to flow of gas	Thermal conductivity
Condensation and reaction	Enthalpy of processes
Electrostatic and magnetic forces	Sensor prearrangement
Electronic drift	Electronic drift

Table 3. Thermal characterization of pharmaceutical bio-based polysaccharides.

Source of polymer	DSC	Tg	TGA[a]/DTG[b]	References
Cassia uniflora	Ed$_p$: 122.5 °C; On$_s$: 112.22 °C; En$_s$: 133.76 °C	186.8 °C and 339.74 °C	-	[12]

(Table 3) cont.....

Source of polymer	DSC	Tg	TGA[a]/DTG[b]	References
Diospyros melonoxylon	Ed_p: 78 °C	74 °C	70.1 °C; 304.7 °C; 475.4 °C	[13]
Buchanania lanzan	Ed_p: 89 °C	80.93 °C – 91.96 °C	352.6 °C; 508.6 °C	[13]
Manilkara zapota	Ed_p: 138 °C	138 °C	350.3 °C; 614.4 °C	[13]
Ocimum basilicum L.	On_s: 146.9 °C; En_s: 148.3 °C	146 °C	-	[14]
Tamarindus indica L.	(Curve-1: Ed_p: 97.67 °C; On_s: 65 °C; En_s: 175 °C) (Curve-2: Ex_p: 310.86 °C; On_s: 245 °C; En_s: 400 °C)	-	70.1 °C; 175.1 °C	[15]
Salvia hispanica	(Curve-1: Ed_p: 97.5 °C; On_s: 84 °C; En_s: 111 °C) (Curve-2: Ed_p: 202 °C; On_s: 190 °C; En_s: 220 °C) (Curve-3: Ex_p: 244 °C; On_s: 230 °C; En_s: 250 °C)	42.93 °C – 57.93 °C	100 °C; 250 °C; 340 °C	[16, 17]
Ornithogalum cuspidatum	Ed_p: 106 °C	7.7 °C, -5.9°C, -8.4°C	100 °C; 230 °C	[18]
Artocarpus heterophyllus	-	Not detected	400 °C;	[19]
Hollyhocks plant seed	-	-	20-129 °C; 130-224 °C; 225-341 °C; 341-417 °C; 418-600 °C;	[20]
Pereskia aculeata Miller	Ed_p: 68 °C; On_s: 61 °C; En_s: 73 °C; Ex_p: 102 °C; Ex_p: 175 °C	- 5 °C	100 °C; 180-400 °C; 350-500 °C;	[21]
Colocasia esculenta	Ed_p: 84.6 °C	-	100 °C; 180-150 °C; 150-500 °C; 272.9-279.1°C[b]	[22, 23]
Opuntia ficus indica	Ed_p: 105 °C; Ex_p: 273 °C	-	50-225 °C; 250-325 °C	[24]
Cassia obtustifolia	Ed_p: 83 °C; Ex_p: 280 °C	193 °C	-	[25]
Okra mucilage	Ex_p: 595.58 °C;	-	120 °C; 271.68 °C; 325.08 °C	[26]
Rice bran starch	Ed_p: 110.14 °C; Ed_p: 316.25 °C; Ex_p: 547.40 °C; Ex_p: 698.14 °C	-	110 °C; 271.68 °C; 385.46 °C	[26]
Sugarcane bagasse	Ed_p: 363.02 °C; Ex_p: 100-330 °C	-	256.78 - 412.02 °C; 359.14 °C	[26]

(Table 3) cont.....

Source of polymer	DSC	Tg	TGA^a/DTG^b	References
Gracilaria corticata	Ed$_p$: 77.88 °C; On$_s$: 62.32 °C; Ex$_p$: 243.30 °C	-	289 °C; 362 °C; 817 °C	[27]
Corchorus olitorius	Ed$_p$: 117.86 °C; On$_s$: 62.25 °C; En$_s$: 132.48 °C	-	-	[28]
Hardwood cellulose and hemicellulose	-	-	300 - 380 °C; 190 - 360 °C	[29]
Sorghum saccharatum	(Curve-1: Ed$_p$: 90.16 °C; On$_s$: 72.65 °C; En$_s$: 89.94 °C); (Curve-2: Ed$_p$: 123.04 °C; On$_s$: 107.51 °C); (Curve-3: Ed$_p$: 204.66 °C; On$_s$: 189.63 °C)	75.6 °C	-	[30]
Dioscorea trifida	Ed$_p$: 90.56 °C; On$_s$: 32.09 °C; En$_s$: 136.24 °C	-	-	[31]
Linum usitatissiumum L.	(Curve-1: Ed$_p$: 73.77 °C; On$_s$: 31.98 °C; En$_s$: 200.57 °C) (Curve-2: Ed$_p$: 267.09 °C; On$_s$: 246.06 °C; En$_s$: 359.53 °C)	-	-	[32]
Cordia lutea	Ed$_p$: 158 °C	-	162.7 °C	[33]
Dillenia indica	-		210 – 500 °C	[34]
Plantago psyllium L	Ed$_p$: 313 °C	-	207 -329 °C; 329 - 600°C	[35]
Borassus flabellifer	Ed$_p$: 292.63 °C	96.06 °C	50 – 400 °C	[36]

DSC: differential scanning calorimetry; TGA^a: mass loss at various temperature during thermogravimetric analysis; Tg: glass transition temperature, Ed$_p$: endothermic peak; Ex$_p$: exothermic peak; On$_s$: oneset of temperature; En$_s$: endset temperature; DTG^b: derivative thermogravimetric peak.

Mechanical Properties of Nanocomposites Fabricated Using Bio-Based Polymers

Bio-based polymers have attracted great interest in recent years for the development of composites, due to their biodegradability and compatibility with remarkable thermomechanical improvement at lower fillers fortification [37]. The word "nanocomposite" was first introduced in 1961 by Blumstein [38], later in 1965, a primitive nanocomposite was investigated for the improvement in thermal stability [39]. The American Society for Testing and Materials (ASTM) has introduced several standard methods for testing the mechanical properties of composites, among them ASTM – D412 has been widely used for the tensile strain-stress determination. ASTM – D412 has replaced 4001, 4116, 4121, and

4411 of Federal Test Standard Method No. 612. Moreover, ASTM – D412 is denoted as the standard method of Tension Testing for Vulcanized Rubber and Rubber-like products [40]. Recently reported mechanical property of composites fabricated using bio-based polymers is presented in Table **4**. In addition, the standard conditions followed during the determination of mechanical property such as Tensile stress (mPa), elongation at break (%), and young's modulus (N/m²) of composite have been elaborated.

Table 4. Mechanical characterization of pharmaceutical bio-based polysaccharides fabricated composites.

Composite Composition	TS	Y	EAB	Applications	References
Morinda citrifolia, glycerol, and *Vaccinium corymbosum* extract	13.0 ± 1.25	131.97 ± 16.4	9.89 ± 0.62	Food packaging with antioxidant potential	[41]
Canna indica L starch with glycerol	1.10 ± 0.6	532.71 ± 0.2	5.93 ± 0.2	Food packaging	[42]
Sodium alginate, *k*-carrageen, and *Momordica cochinchinensis*	1241 ± 108	14797 ± 3040	18.4 ± 2.5	Food coating	[43]
Pectin and glycerol	14.31 ± 0.6	-	1.20 ± 0.03	Food packaging	[44]
Gum karaya, glycerol, and cloisite	5.44 ± 0.89	-	12.68 ± 2.1	Food packaging	
Bovine gelatin, glycerol, peptides,	Blend film: 85.5 ± 3.9; Bilayer film 59.88 ± 8.0	-	Blend film: 11.34 ± 1.9; Bilayer film 26.42 ± 3.7	Food packaging	[45]
Kappaphycus alvarezii, glycerol	44.63 ± 3.95	-	3.80 ± 0.52	Food packaging	[46]
Fish skin gelatin, agar, glycerol	1.77 ± 0.14; 18.69 ± 3.9	9.05 ± 3.52; 361.6 ± 44.87	76.73 ± 17.3; 361.67 ± 44	Food packaging	[47]
Cashew gum, polyvinyl alcohol	23.7 ± 2.8	-	187.2 ± 9.3	Antifungal film	[48]
Exopolysaccharide produced by *Pseudomonas oleovorans*	51.0 ± 3.0	1738 ± 114	9.5 ± 3.9	Nanocomposite	[49]
k-carrageen, locust bean gum, Cloisite 30B	33.82 ± 0.51	-	29.82 ± 1.01	Food packaging	[50]
Pereskia aculeata Miller, glycerol	2.89 ± 0.03	64.01 ± 0.78	27.79 ± 0.59	Edible packaging film	[51]
Chitosan, polyvinyl alcohol, glycerol, anthocyanin	14.72 ± 0.13; 16.67 ± 0.06	0.04 ± 0.08; 0.03 ±0.005	227.41 ± 0.27; 167.17 ± 0.87	Beverage freshness	[52]

(Table 4) cont.....

Composite Composition	TS	Y	EAB	Applications	References
Chitosan, polyvinyl alcohol, glycerol, silver nanoparticles	20.0 ± 1.0	0.039 ± 0.006	52.0 ± 1.0	Food packaging	[53]
Chitosan, anthocyanin	3.50 ± 0.434	0.048 ± 0.007	65.0 ± 7.48	Food preservation	[54]

TS: Tensile strength(MPa); **Y:** Young's modulus (MPa); **EAB:** elongation at break (%).

- Sample preparation for the testing of mechanical strength includes either dumbbell (Test A) or standard ring (Test B) that must be flat in shape.
- The overall length of the dumbbell shape should not exceed 115 mm with the width of 6 mm and not less than 1.5 mm or more than 3 mm thickness.
- Micrometer must be capable of exerting a pressure of 25.8 ± 4.8 kPa with a maximum film thickness of 0.001 mm.
- The tester instruments must have two grips, one of which shall be connected to the dynamometer, and a mechanism for separating the grips at a uniform rate of 500 ± 50 mm/min for a distance of at least 75 mm or until the specimen break.

CONCLUSIONS AND FUTURE ASPECTS

The quantification of thermomechanical properties of polymer demonstrates an important role in the development of a novel drug delivery system. As formulation becomes complex and its characterization becomes more difficult, industrial manufacturers are keeping up the pace with innovation of more particular, sensitive, and durable instruments. This chapter presents recent and emerging improvements in the pharmaceutical field, whose future applications are limited only by the investigators' imagination and budget.

REFERENCES

[1]   Mildner SA, Lauster G, Wodni W. Scarcity and abundance revisited: A literature review on natural resources and conflict. Int J Confl Violence 2011; 5(1): 155-72.

[2]   Roy BC, Sarkar S, Mandal NR. Natural resource abundance and economic Performance: a literature review. Current Urban Studies 2013; 1(4): 148-55.
[http://dx.doi.org/10.4236/cus.2013.14016]

[3]   Navarro DMDL, Abelilla JJ, Stein HH. Structures and characteristics of carbohydrates in diets fed to pigs: a review. J Anim Sci Biotechnol 2019; 10(1): 39.
[http://dx.doi.org/10.1186/s40104-019-0345-6] [PMID: 31049199]

[4]   Engelking LR. Chemical composition of living cells. In: Engelking LR, Ed. Textbook of veterinary physiological chemistry. 3rd ed. Boston: Academic Press 2015; pp. 2-6.
[http://dx.doi.org/10.1016/B978-0-12-391909-0.50001-3]

[5]   Zheng Y, Xie Q, Wang H, Hu Y, Ren B, Li X. Recent advances in plant polysaccharide-mediated nano drug delivery systems. Int J Biol Macromol 2020; 165(Pt B): 2668-83.
[http://dx.doi.org/10.1016/j.ijbiomac.2020.10.173] [PMID: 33115646]

[6]   Stodghill SP. Thermal analysis: a review of techniques and applications in the pharmaceutical sciences. Americ Pharm Rev 2010; 13(12)

[7]   Schmitt EA, Peck K, Sun Y, Geoffroy JM. Rapid, practical and predictive excipient compatibility screening using isothermal microcalorimetry. Thermochim Acta 2001; 380(2): 175-84.
[http://dx.doi.org/10.1016/S0040-6031(01)00668-2]

[8]   Haines PJ, Reading M, Wilburn FW. Differential Thermal Analysis and Differential Scanning Calorimetry. Handbook of thermal analysis and calorimetry. Elsevier Science BV 1998; pp. 279-361.
[http://dx.doi.org/10.1016/S1573-4374(98)80008-3]

[9]   Joseph D, Menczel LJ, Bruce R Prime, Harvey E Bair, Mike Reading, Steven Swier. Differential scanning calorimetry (DSC). In: Joseph D, Menczel BRP, Eds. Thermal analysis of polymer: fundamentals and applications: Wiley. 1999; pp. 7-8.

[10]   Earnest CM. Compositional analysis by thermogravimetry ASTM STP 997-EB. Philadelphia: ASTM International 1988; p. 298.
[http://dx.doi.org/10.1520/STP997-EB]

[11]   Nwabor OF, Singh S, Wunnoo S, Lerwittayanon K, Voravuthikunchai SP. Facile deposition of biogenic silver nanoparticles on porous alumina discs, an efficient antimicrobial, antibiofilm, and antifouling strategy for functional contact surfaces. Biofouling 2021; 37(5): 538-54.
[http://dx.doi.org/10.1080/08927014.2021.1934457] [PMID: 34148443]

[12]   Deore UV, Mahajan HS. Isolation and structural characterization of mucilaginous polysaccharides obtained from the seeds of *Cassia uniflora* for industrial application. Food Chem 2021; 351: 129262.
[http://dx.doi.org/10.1016/j.foodchem.2021.129262] [PMID: 33626466]

[13]   Bothara SB, Singh S. Thermal studies on natural polysaccharide. Asian Pac J Trop Biomed 2012; 2(2): S1031-5.
[http://dx.doi.org/10.1016/S2221-1691(12)60356-6]

[14]   Nazir S, Wani IA. Functional characterization of basil (*Ocimum basilicum* L.) seed mucilage. Bioactive Carbohydrates and Dietary Fibre 2021; 25: 100261.
[http://dx.doi.org/10.1016/j.bcdf.2021.100261]

[15]   Alpizar-Reyes E, Carrillo-Navas H, Gallardo-Rivera R, Varela-Guerrero V, Alvarez-Ramirez J, Pérez-Alonso C. Functional properties and physicochemical characteristics of tamarind (*Tamarindus indica* L.) seed mucilage powder as a novel hydrocolloid. J Food Eng 2017; 209: 68-75.
[http://dx.doi.org/10.1016/j.jfoodeng.2017.04.021]

[16]   Timilsena YP, Adhikari R, Kasapis S, Adhikari B. Molecular and functional characteristics of purified gum from *Australian chia* seeds. Carbohydr Polym 2016; 136: 128-36.
[http://dx.doi.org/10.1016/j.carbpol.2015.09.035] [PMID: 26572338]

[17]   Velázquez-Gutiérrez SK, Figueira AC, Rodríguez-Huezo ME, Román-Guerrero A, Carrillo-Navas H, Pérez-Alonso C. Sorption isotherms, thermodynamic properties and glass transition temperature of mucilage extracted from chia seeds (Salvia hispanica L.). Carbohydr Polym 2015; 121: 411-9.
[http://dx.doi.org/10.1016/j.carbpol.2014.11.068] [PMID: 25659716]

[18]   Gheybi N, Pirouzifard MK, Almasi H. *Ornithogalum cuspidatum* mucilage as a new source of plant-based polysaccharide: Physicochemical and rheological characterization. J Food Meas Charact 2021; 15(3): 2184-201.
[http://dx.doi.org/10.1007/s11694-021-00814-z]

[19]   Prakash A, Lata R, Martens PJ, Rohindra D. Characterization and *in-vitro* analysis of poly(ε-caprolactone)-"Jackfruit" Mucilage blends for tissue engineering applications. Polym Eng Sci 2021; 61(2): 526-37.
[http://dx.doi.org/10.1002/pen.25597]

[20]   Nowrouzi I, Mohammadi AH, Khaksar Manshad A. Characterization and likelihood application of extracted mucilage from Hollyhocks plant as a natural polymer in enhanced oil recovery process by alkali-surfactant-polymer (ASP) slug injection into sandstone oil reservoirs. J Mol Liq 2020; 320: 114445.

[http://dx.doi.org/10.1016/j.molliq.2020.114445]

[21]   Silva SH, Neves ICO, Oliveira NL, *et al.* Extraction processes and characterization of the mucilage obtained from green fruits of Pereskia aculeata Miller. Ind Crops Prod 2019; 140: 111716.
[http://dx.doi.org/10.1016/j.indcrop.2019.111716]

[22]   Mijinyawa AH, Durga G, Mishra A. Evaluation of thermal degradation and melt crystallization behavior of taro mucilage and its graft copolymer with poly(lactide). SN Applied Sciences 2019; 1(11): 1486.
[http://dx.doi.org/10.1007/s42452-019-1490-4]

[23]   Mijinyawa AH, Durga G, Mishra A. Isolation, characterization, and microwave assisted surface modification of *Colocasia esculenta* (L.) Schott mucilage by grafting polylactide. Int J Biol Macromol 2018; 119: 1090-7.
[http://dx.doi.org/10.1016/j.ijbiomac.2018.08.045] [PMID: 30099042]

[24]   Otálora MC, Gómez Castaño JA, Wilches-Torres A. Preparation, study and characterization of complex coacervates formed between gelatin and cactus mucilage extracted from cladodes of Opuntia ficus-indica. Lebensm Wiss Technol 2019; 112: 108234.
[http://dx.doi.org/10.1016/j.lwt.2019.06.001]

[25]   Deore UV, Mahajan HS. Isolation and characterization of natural polysaccharide from *Cassia Obtustifolia* seed mucilage as film forming material for drug delivery. Int J Biol Macromol 2018; 115: 1071-8.
[http://dx.doi.org/10.1016/j.ijbiomac.2018.04.174] [PMID: 29727659]

[26]   Chandra Mohan C, Harini K, Vajiha Aafrin B, *et al.* Extraction and characterization of polysaccharides from tamarind seeds, rice mill residue, okra waste and sugarcane bagasse for its Bio-thermoplastic properties. Carbohydr Polym 2018; 186: 394-401.
[http://dx.doi.org/10.1016/j.carbpol.2018.01.057] [PMID: 29456002]

[27]   Archana G, Sabina K, Babuskin S, *et al.* Preparation and characterization of mucilage polysaccharide for biomedical applications. Carbohydr Polym 2013; 98(1): 89-94.
[http://dx.doi.org/10.1016/j.carbpol.2013.04.062] [PMID: 23987320]

[28]   Azubuike C, Alfa M, Oseni B. Characterization and evaluation of the suspending potentials of *Corchorus olitorius* mucilage in pharmaceutical suspensions. Tropical Journal of Natural Product Reseach 2017; 1(1): 39-46.
[http://dx.doi.org/10.26538/tjnpr/v1i1.7]

[29]   Shen DK, Gu S, Bridgwater AV. The thermal performance of the polysaccharides extracted from hardwood: Cellulose and hemicellulose. Carbohydr Polym 2010; 82(1): 39-45.
[http://dx.doi.org/10.1016/j.carbpol.2010.04.018]

[30]   Rivera-Corona JL, Rodríguez-González F, Rendón-Villalobos R, García-Hernández E, Solorza-Feria J. Thermal, structural and rheological properties of sorghum starch with cactus mucilage addition. Lebensm Wiss Technol 2014; 59(2): 806-12.
[http://dx.doi.org/10.1016/j.lwt.2014.06.011]

[31]   Pires MB, Amante ER, da Cruz Rodrigues AM, da Silva LHM. Isolation and characterization of starch from purple yam (Dioscorea trifida). J Food Sci Technol 2021.

[32]   Kaur M, Kaur R, Punia S. Characterization of mucilages extracted from different flaxseed (*Linum usitatissiumum* L.) cultivars: A heteropolysaccharide with desirable functional and rheological properties. Int J Biol Macromol 2018; 117: 919-27.
[http://dx.doi.org/10.1016/j.ijbiomac.2018.06.010] [PMID: 29874558]

[33]   Troncoso OP, Zamora B, Torres FG. Thermal and rheological properties of the mucilage from the fruit of *Cordia lutea*. Polymers from Renewable Resources 2017; 8(3): 79-90.
[http://dx.doi.org/10.1177/204124791700800301]

[34]   Bal T, Murthy PN, Sengupta S. Isolation and analytical studies of mucilage obtained from the seeds of

Dillenia indica (family dilleniaceae) by use of various analytical techniques. Asian J Pharm Clin Res 2012; 5(3): 65-8.

[35]   Chinellato MM. Properties of mucilage blends using psyllium husk (Plantago psyllium L) and chia seed (Salvia hispanica L). A PhD dissertation State Univeristy of Maringa 2021.

[36]   Ravi Kumar RN, Narayana Swmay VB. Isolation and Characterization of Borassus flabellifer Mucilage. Research J Pharm and Tech 2012; 5(8): 1093-101.

[37]   Zhang B, Liang Y, Liu B, Liu W, Liu Z. Enhancing the Thermo-Mechanical Property of Polymer by Weaving and Mixing High Length–Diameter Ratio Filler. Polymers (Basel) 2020; 12(6): 1255.
[http://dx.doi.org/10.3390/polym12061255] [PMID: 32486186]

[38]   Blumstein A. Etude des polymerisations en couche adsorbee. Bull Soc Chim Fr 1961; (5): 899-906.

[39]   Blumstein A. Polymerization of adsorbed monolayers. II. Thermal degradation of the inserted polymer. J Polym Sci A 1965; 3(7): 2665-72.
[http://dx.doi.org/10.1002/pol.1965.100030721]

[40]   Standard test methods for vulcanized rubber and thermoplastic elastomers Tension. United States of America 2013; pp. 1-13.

[41]   Han HS, Song KB. Noni (*Morinda citrifolia*) fruit polysaccharide films containing blueberry (*Vaccinium corymbosum*) leaf extract as an antioxidant packaging material. Food Hydrocoll 2021; 112: 106372.
[http://dx.doi.org/10.1016/j.foodhyd.2020.106372]

[42]   Cabrera Canales ZE, Rodríguez Marín ML, Gómez Aldapa CA, Méndez Montealvo G, Chávez Gutiérrez M, Velazcuez G. Effect of dual chemical modification on the properties of biodegradable films from achira starch. J Appl Polym Sci 2020; 137(45): 49411.
[http://dx.doi.org/10.1002/app.49411]

[43]   Tran TTB, Roach P, Nguyen MH, Pristijono P, Vuong QV. Development of biodegradable films based on seaweed polysaccharides and Gac pulp (*Momordica cochinchinensis*), the waste generated from Gac oil production. Food Hydrocoll 2020; 99: 105322.
[http://dx.doi.org/10.1016/j.foodhyd.2019.105322]

[44]   Gouveia TIA, Biernacki K, Castro MCR, Gonçalves MP, Souza HKS. A new approach to develop biodegradable films based on thermoplastic pectin. Food Hydrocoll 2019; 97: 105175.
[http://dx.doi.org/10.1016/j.foodhyd.2019.105175]

[45]   Abdelhedi O, Nasri R, Jridi M, *et al.* Composite bioactive films based on smooth-hound viscera proteins and gelatin: Physicochemical characterization and antioxidant properties. Food Hydrocoll 2018; 74: 176-86.
[http://dx.doi.org/10.1016/j.foodhyd.2017.08.006]

[46]   Farhan A, Hani NM. Characterization of edible packaging films based on semi-refined kappa-carrageenan plasticized with glycerol and sorbitol. Food Hydrocoll 2017; 64: 48-58.
[http://dx.doi.org/10.1016/j.foodhyd.2016.10.034]

[47]   Mohajer S, Rezaei M, Hosseini SF. Physico-chemical and microstructural properties of fish gelatin/agar bio-based blend films. Carbohydr Polym 2017; 157: 784-93.
[http://dx.doi.org/10.1016/j.carbpol.2016.10.061] [PMID: 27987991]

[48]   Barbara DSS, Cirano JU, Karla AB, Maria C, Di M. Romulo Roosevelt da SF, Fabio Y, Katia FF. Biodegradable and bioactive CGP PVA film for fungal growth inhibition. Carbohydr Polym 2012; 89(3): 964-70.
[http://dx.doi.org/10.1016/j.carbpol.2012.04.052] [PMID: 24750887]

[49]   Alves VD, Ferreira AR, Costa N, Freitas F, Reis MAM, Coelhoso IM. Characterization of biodegradable films from the extracellular polysaccharide produced by Pseudomonas oleovorans grown on glycerol byproduct. Carbohydr Polym 2011; 83(4): 1582-90.
[http://dx.doi.org/10.1016/j.carbpol.2010.10.010]

[50]   Martins JT, Bourbon AI, Pinheiro AC, Souza BWS, Cerqueira MA, Vicente AA. Biocomposite films based on κ-carrageenan/locust bean gum blends and clays: physical and antimicrobial properties. Food Bioprocess Technol 2013; 6(8): 2081-92.
[http://dx.doi.org/10.1007/s11947-012-0851-4]

[51]   Oliveira NL, Rodrigues AA, Oliveira Neves IC, Teixeira Lago AM, Borges SV, de Resende JV. Development and characterization of biodegradable films based on Pereskia aculeata Miller mucilage. Ind Crops Prod 2019; 130: 499-510.
[http://dx.doi.org/10.1016/j.indcrop.2019.01.014]

[52]   Singh S, Nwabor OF, Syukri DM, Voravuthikunchai SP. Chitosan-poly(vinyl alcohol) intelligent films fortified with anthocyanins isolated from *Clitoria ternatea* and *Carissa carandas* for monitoring beverage freshness. Int J Biol Macromol 2021; 182: 1015-25.
[http://dx.doi.org/10.1016/j.ijbiomac.2021.04.027] [PMID: 33839180]

[53]   Nwabor OF, Singh S, Paosen S, Vongkamjan K, Voravuthikunchai SP. Enhancement of food shelf life with polyvinyl alcohol-chitosan nanocomposite films from bioactive Eucalyptus leaf extracts. Food Biosci 2020; 36: 100609.
[http://dx.doi.org/10.1016/j.fbio.2020.100609]

[54]   Eze FN, Jayeoye TJ, Singh S. Fabrication of intelligent pH-sensing films with antioxidant potential for monitoring shrimp freshness *via* the fortification of chitosan matrix with broken Riceberry phenolic extract. Food Chem 2022; 366: 130574.
[http://dx.doi.org/10.1016/j.foodchem.2021.130574] [PMID: 34303209]

CHAPTER 6

Pharmaceutical and Biomedical Applications of Bio-Based Excipients

Abstract: The success of an active pharmaceutical depends on how efficiently and precisely the polymeric dosage form can deliver it for effective treatment. Polymers are recognized as inactive pharmaceutical excipients and the backbone of the drug delivery system that plays an essential role in the design of dosage forms. Biodegradable polymer-based drug delivery system has gained significant attention among researchers and manufacturers in the last few decades, compared to synthetic non-biodegradable and their analog polymers. Synthetic biodegradable biopolymers demonstrate excellent efficacy in the design and development of drug delivery that enables the incorporation of active pharmaceuticals into the body. Despite the wide effectiveness of currently available polymers in the design of drug delivery systems, the quest for biocompatible, biodegradable, and easily accessible novel polymers with multifarious applications is still protractile. Due to safety and regulatory approval requirements in the development of novel inactive pharmaceuticals, the introduction of new excipients is much limited. However, the development of bio-based polymers with modification as required could be a valuable way to address the problem associated with synthetic polymers. In this chapter, an overview has been presented on the various applications of bio-based polymers ranging from oral conventional drug delivery to reduction and capping of metallic materials. Moreover, details are presented on the technology-based use of biopolymers in the fabrication of modified oral drug delivery, microneedles, packaging film, and biogenic synthesis of metallic nanoparticles.

Keywords: Bio-based polymers, Biogenic metallic nanoparticles, Electro-spinning, Modified drug delivery, Tissue engineering.

INTRODUCTION

Today polymer industry has grown significantly due to its wide applicability in various sectors. The manufacturing and output of polymers per unit volume have left behind several other industrial materials such as aluminum, copper, steel, and other major metallic materials [1]. Polymers are a large structural linkage of the monomeric unit and frequently form a chain-like arrangement. The application of polymers has widened day by day, compared to other classes of materials developed by humans. The first man-made polymers have been studied since 1832 [2]. The contemporary applications of polymers extend from a

Sudarshan Singh & Warangkana Chunglok

pharmaceutical dosage form to packaging materials, composites, textile, coating, foams, adhesives, electronic devices, optical devices, industrial fibers, and the development of advanced ceramics (Fig. **1**).

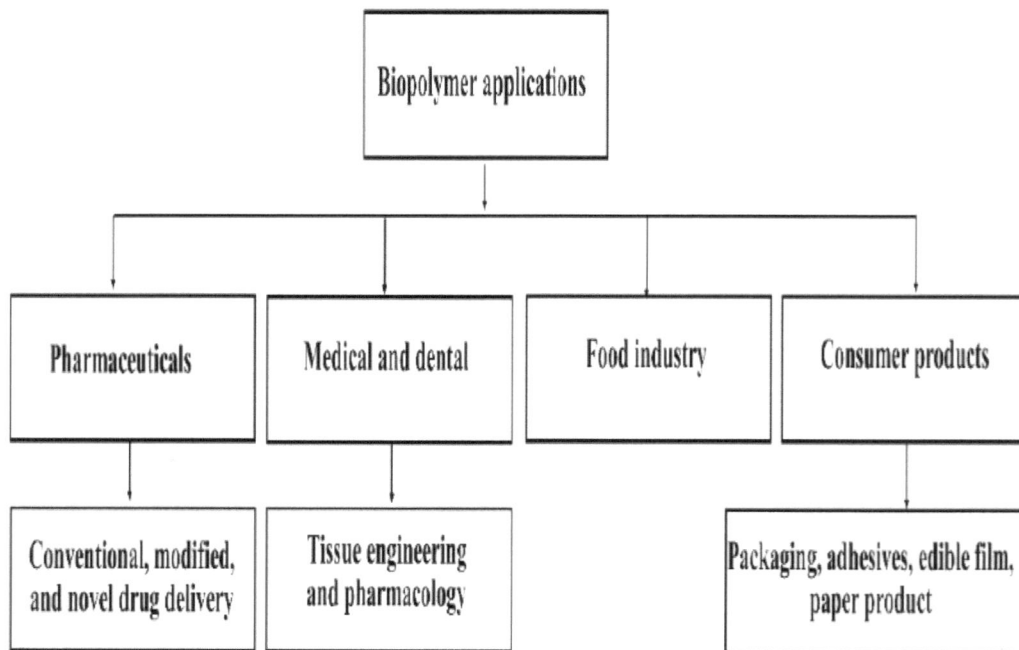

Fig. (1). Application of bio-based polymers in various fields.

Approximately 140 million tons of synthetic polymers are produced around the globe annually for various applications by the manufacturer [1]. Most synthetic polymers and integrated polymer products are stable and require an unlimited time for degradation. Moreover, due to stability and resistance to degradation of synthetic polymer products, they accumulate in the environment at a rate of 8% by weight and 20% by volume in landfills, which causes a serious impact on the ecosystem [3]. Although the acceptance of synthetic polymers and intended products is higher than biopolymers, still bio-based polymers are gaining significant attention among researchers and manufacturers for extending application in various sectors. Bio-based polymers are natural polymers produced or synthesized from animals, plants, bacteria, and fungi. In addition, biopolymers could also be developed by a chemical reaction among biological materials such as sugars, amino acids, oils, or natural fats [4]. Analogous to synthetic biopolymers excipients, bio-based polymers consist of monomeric units covalently bonded to form a larger structure. Perhaps, several limitations are associated with bio-based polymers but research has indicated potential applications in various drug delivery systems and allied areas. In this chapter, a brief account is presented on several applications of polymers in conventional, modified, and novel drug delivery, fabrication of nanofibers and scaffolds using electrospinning techniques, development of microneedles, and in the bio-reduction of metallic nanoparticles.

Polymers in Conventional Drug Delivery System

Several routes for the administration of drugs are available such as oral, mucosal, parenteral, transdermal, pulmonary, *etc*. Among these delivery routes for a therapeutic agent, oral route is widely accepted. Oral formulations have been broadly classified based on the local or systemic efficacy as a conventional and modified or novel drug delivery system. Recent advancements in drug delivery and polymer chemistry shifted the fabrication technology from conventional to modified drug delivery systems with an enhancement in bioavailability and therapeutic efficacy. However, conventional drug delivery is still the most promising drug administration route, due to several advantages including the ease of self-administration, non-invasiveness, patient preference, and low cost-o--production [5]. The conventional drug delivery system has been the most accepted delivery system among pharmaceutical manufacturers and patients. However, drugs with poor solubility, stability, and permeability exhibit low bioavailability and pharmacological efficacy when administered orally [6]. Fabrication and development of oral drug delivery system require the addition of inactive materials such as additives, binder, diluents, lubricants, glidants, *etc*. to produce a suitable dosage form with the effective release of active pharmaceuticals. These inactive materials are denoted as excipients derived from either synthetic or natural sources. Although synthetic polymers have been effectively used in the development of oral conventional and novel drug delivery systems, however biodegradability is still of concern. Whereas, natural polymers are gaining interest among scientists and manufacturers due to their excellent biodegradability within nature.

Bio-based natural polymers have been extensively used as a tablet binder and explored in the design of various oral drug delivery systems including fast disintegrating, orodispersible, bio and mucoadhesive tablets, *etc*. In addition, bio-based polymers have been tested successfully in the fabrication of several novel drug delivery systems such as microsphere, nanoparticles, microbeads, transdermal films, wound dressing scaffolds, thermo-responsive gel, continuous or pulsatile release systems, *etc*. Moreover, advancements in pharmaceutical technology with a modification in polymer chemistry have significantly changed the formulation aspects of conventional drug delivery systems. Several conventional oral solid dosage forms incorporated with synthetic polymers have been effectively modified to enhance the systemic efficacy of active pharmaceuticals using various techniques and patented as WOWTAB®, ORASOLV®, EFVDAS®, FLASHTAB®, ZYDIS®, LYOC®, QUICKSOLV®, and FLASHDOSE®. Similarly, bio-based polymers obtained from *Ocimum basilicum* [8, 9], *Trigonella foenum-graecum* [8, 10], *Plantago ovata* [11 - 13], *Salvia hispanica* [14], and *Hibiscus rosa sinensis* [13, 15] have also been reported in the development of fast disintegrating oral solid dosage form. While polymers obtained from *Hibiscus rosa sinensis* [16], *Eulophia herbacae* [17], *Remusatia vivipara* [17], *Artocarpus heterophyllus* [18], *Basella alba* [19], B*rachystegia eurcoma* [20], *Delonix regia* [21], *Mimosa pudica* [22], and *Mulva neglecta* [23] have been reported as excellent tablet binders, emulsifiers, suspending agents, and release modifiers in an oral drug delivery dosage form. Recent reports on the bio-based polymer used for conventional oral and modified drug delivery system has been presented in Table **1**.

Table 1. Potential application of bio-based polymers in the fabrication of oral conventional, modified, and novel drug delivery systems.

Biopolymer	Delivery System	Application	Reference
Ocimum basilicum	Magnetic nano-hydrogel	Superabsorbent magnetic nano-hydrogel of Naproxen prepared *via* free radical copolymerization between vinyl modified Fe_3O_4/SiO_2 nanoparticles and natural polymer demonstrated excellent absorption phenomena with potential as a local delivery application.	[39]
Chitosan	Sponge composite	Chitosan lyophilized sponge with *Hibiscus syriacus* demonstrated promising and potential wound healing effects.	[40]
Psyllium husk	Microparticles	Isabgoal husk loaded gliclazide microparticles showed improved oral bioavailability with prolonged hypoglycemic effects.	[41]
Linum usitatissium **L**	Ternary conjugates	Flaxseed mucilage complexed with oxidized tannic acid showed potential application in the controlled release of bioactive.	[42]
Psyllium husk	Mucoadhesive microsphere	Cross-linked gliclazide loaded mucoadhesive microsphere showed sustained release of active compound for control of diabetic hyperglycemia.	[43]
Mimosa pudica **chitosan and alginate**	Float-adhesive microsphere	Microsphere with the dual advantage of floating and mucoadhesion showed potential towards improvement in oral bioavailability.	[44]
Hibiscus rosa-sinensis **L.**	Transdermal drug delivery	Leave mucilage showed enhanced transdermal permeation of caffeine.	[45]

(Table 1) cont.....

Biopolymer	Delivery System	Application	Reference
Plantago ovata	Electrospun scaffold	Mucilage combined with polyvinyl alcohol nanofibers showed promising scaffold application in the dermal drug delivery system.	[46]
Ocimum basilicum L	Wound dressing	pH-sensitive hydrogel films based on a combination of natural polymer and polyvinyl alcohol demonstrated biocompatibility with a controlled release of drug .	[47]
Ocimum basilicum L	Beads	A combination of sodium alginate and basil seed mucilage showed excellent release retardant with improvement in the stability of metformin.	[48]
Trigonella foenum-graecum	Scaffold	Biodegradable microwave-assisted polyacrylamide graft blend with mucilage-polyvinyl alcohol showed sufficient tissue growth with controlled release of the drug.	[49]
Lallemantia royaleana Benth	Niosomes	Combination of mucilage with carbopol in the different ratios indicated enhancement in permeability with prolonged duration of action for topical gel containing nisomes.	[50]
Colocasia esculenta	Microsphere	Mucilage alginate microsphere indicated improved bioavailability of pregabalin.	[51]
Aloe vera	Encapsulation	Spray-dried curcumin encapsulated using mucilage demonstrated excellent release retardant efficacy with potential application in pharmaceuticals.	[52, 53]
Anacardium occidentale	Dental paste	Aceclofenac dental paste comprised of natural gum demonstrated potential in effective management of dental inflammation and pain.	[54]
Diospyros melonoxylon Roxb	Oral mucoadhesive	Isolated mucilage demonstrated good mucoadhesion as well as controlled release of drug.	[31, 55]
Buchanania lanzan spreng	Oral mucoadhesive	Biodegradable mucilage suggested potential application in the mucoadhesive delivery system.	[56 - 58]
Annona squamosa	Buccal bioadhesive	Natural polysaccharides demonstrated good bioadhesive strength compared to carbopol and polycarbophil.	[59]
Pithecellobium dulce	Oral mucoadhesive	Isolated mucilage demonstrated notable bioadhesive strength compared to xanthan gum and hydroxyl propyl methyl cellulose K4.	[60]
Manilkara zapota	Oral mucoadhesive	Seed mucilage of *Manilkara zapota* indicated effective usage in wet granulation technology with mucoadhesion compared to hydroxyl propyl methyl cellulose E5LV.	[61 - 64]

(Table 1) cont.....

Biopolymer	Delivery System	Application	Reference
Abelmoschus esculentus	Microsphere	*In vitro* release profile suggested that okra pod mucilage could be potential pharmaceutical excipients.	[65]
Hibiscus esculantus	Buccal film	Blend of chitosan with okra mucilage buccal film imparted good mechanical and control diffusion of the drug.	[66]
Colocasia esculenta	Microsphere	Release profile and encapsulation efficacy demonstrated that mucilage could be an effective polymer in the design of microsphere.	[67]
Caesalpinia pulchirrima, Leucaena leucocephala	Oral mucoadhesive	Isolated mucilage demonstrated mucoadhesion similar to hydroxyl propyl methyl cellulose with modulation in the release of diclofenac sodium.	[68]
Aloe vera	Buccal bioadhesive	Buccal bioadhesive tablets using mucilage demonstrated good swelling index and comparative mucoadhesion against guar gum and hydroxyl propyl cellulose.	[69]
Datura stramonium	Bioadhesive film	A blend of mucilage with ethyl cellulose indicated bioadhesive film could suppress the hepatic metabolism of aceclofenac.	[29]

Polymers in Modified And Novel Drug Delivery Systems

Modification in the existing conventional dosage form or incorporation of newly developed excipients that release the active pharmaceuticals in a regulated manner is generally considered a modified drug delivery system. The novel drug delivery system refers to the approaches, formulations, technologies, and systems that deliver the active pharmaceuticals to the desired organ or tissue with required therapeutic efficacy at a lower dose. Novel drug delivery systems especially with polymeric nanocarriers that overcome the physiochemical and pharmacokinetic limitations of pharmaceuticals have gained significant attention among pharmaceutical manufacturers. Nanocarriers improve the therapeutic efficacy of drugs, as well as provide a preferential accumulation at the targeted site. Alterations in the polymeric nanocarrier properties such as constituents, size, shape, and surface charges *via* grafting, spray drying, co-processing, or co-crystallization, *etc.*, are considered a prominent approach to enhancing the efficacy of carrier-based drug delivery systems [24]. Carrier-based drug delivery systems such as micelles, vesicle, liquid crystals dispersion, and nanoparticles have demonstrated a prominent effect in the treatment and management of various diseases including cancer, diabetes, obstructive pulmonary disorder, lower respiratory infections, *etc.* In addition, bio-based polymers and nanocarriers originating from polysaccharides and proteins have been significantly explored in the development of various other novel drug deliveries such as colon targeted, transdermal, periodontal, floating gastro retentive, and mucoadhesive dosage forms.

Gastro-retentive oral drug delivery systems could prolong the residence time of the dosage form and thereby increase the bioavailability of incorporated pharmaceuticals. Several approaches have been adopted by the researchers to increase gastric residence time including the use of excipients that change the density of the dosage form and allow it either to float or be entrapped with mucosa within the gastrointestinal tract [25]. Mucoadhesive drug delivery system refers to the adhesion of a dosage form at the mucus layer and mucin molecules of epithelial intestinal mucosa to increase the residence time for prolonging the action. Mucoadhesion of dosage forms generally arises by high gel-forming hydrophilic polymers derived from either synthetic or natural resources. Synthetic polymers such as hypromellose [26], hydroxypropyl cellulose [27], carboxymethylcellulose [28], sodium alginate [26], *etc.* while polysaccharides originating from *Datura stramonium* leaves [29], *Eulophia herbacea* [30], *Diospyros melonoxylon* [31], *Ziziphus mauritiana* [32], *Buchanania lanzan* [33], *Manilkara zapota* [34], *etc.* are reported as biodegradable natural mucoadhesive polymers.

Polymeric Scaffolds for Tissue Engineering

Injuries and organ failure require immediate attention with an application of either implants or organ transplantations, however due to lack of required numbers, organs are still a matter of concern among clinical surgeons. The current approach to counter this the problem is to rely on specific techniques such as autotransplantation, xeno-transplantation, and implantation of an artificial mechanical organ. Though these techniques provide a satisfactory result, however the cost of fabrication is high. Moreover, this technique might suffer from potential perils such as immunological rejection and viral transmissions with lamination in donor transplantations [35]. Although artificial organ implantation in clinical treatments often results in successful transplantations and improvement in the quality of life,; however, the cost of availing the organ and implants to fulfill the present needs is extremely high. Therefore, polymeric three-dimensional scaffolds for tissue engineering could be a potential alternative to solve the limitation associated with the transplantations and fullfill the urgent need of patients.

Polymeric scaffolds for tissue engineering are a combination of engineering and natural science technology that involves the development of biological materials to replace, repair, and improve the tissue regeneration process [36]. Bio-based polymeric materials offer an excellent opportunity for sustainable development of economically and ecologically attractive materials that could be significantly utilized in the fabrication of dermal nano-fiber and dressings. Moreover, a significant transformation from conventional scaffold fabrication to modern techniques such as electrospun fibers and three-dimensional (3D) printing technology using bio-based polymeric has been reported recently [37]. Moreover, progress in 3D printing is also referred to as additive manufacturing in which complex structures are fabricated especially for tissue engineering. This emerging technology has potential in the fabrication of biodegradable scaffolds with intricate design and structure.

Biocompatibility and biodegradability are the prerequisites for the successful development of scaffolds using polymeric materials. Several materials have been tested for the development of composites and scaffolds including polymers, ceramics, metals, *etc.* Among these, polymers such as synthetic and natural polymers both have been reported in the fabrication of a composite. However, concerning biodegradability and other compatibility-related issues, synthetics are less preferred. While the scaffold prepared using bio-based polymers produces excellent porous shape and mechanical properties, with the ability to support cell growth [38]. Scaffolds for tissue regeneration are prepared basically by two methods such as conventional and novel technology. Conventional methods include solvent casting and thermal-induced phase separation, whereas electrospinning, molding techniques, and 3D printing are advanced technologies as presented in Fig. (**2**).

Tissue engineering (TE) has shown tremendous scope that could promote sustainable enhancement in the quality of human life, with a reduction in healthcare costs. The tissue engineering process demonstrated application in the construction of bio-artificial tissue and *in vitro* and *in vivo* alteration of cell growth. Moreover, TE also functions *via* the implantation of suitable cells isolated from donors. Tissue engineering conglomerates the principles of engineering and biology with the involvement of biomaterials, cells, and bioactive molecules [70]. The approach was initially established to resolve the organ transplantation issue that arises due to its failure and the limited number of donated organs available. Biomaterials used for tissue engineering demand specific property including porosity, controlled surface physics, and chemistry [71]. Moreover, an excellent biodegradability of material is a prerequisite to promote optimal cell adhesion, migration, and deposition of endogenous extracellular matrix materials by cells. Recent reports on bio-based polymeric scaffolds for tissue engineering are presented in Table **2**.

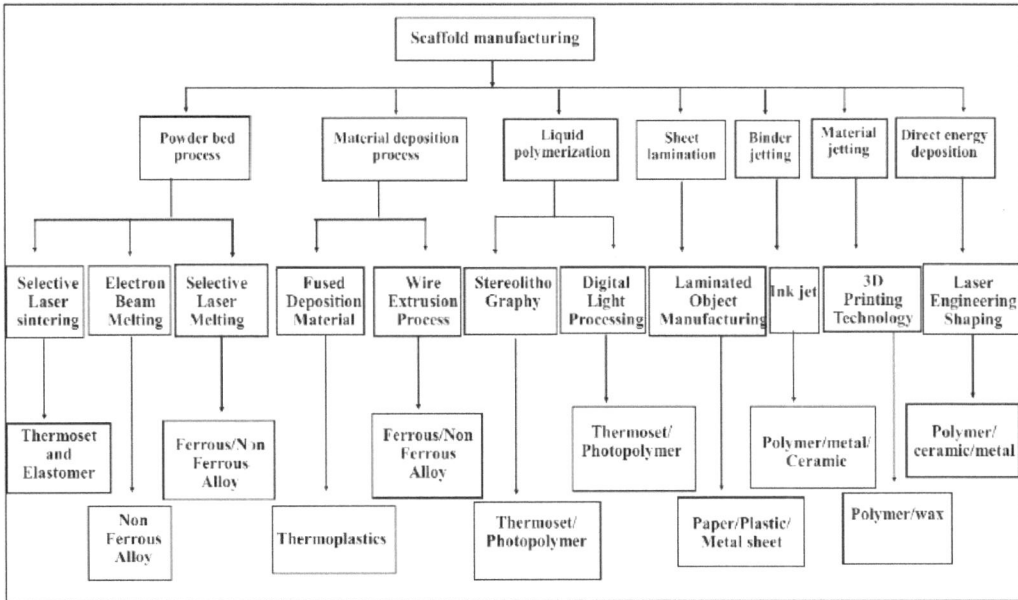

Fig. (2). Classification of various manufacturing processes of the scaffold.

Table 2. Recent scientific reports on the development of biopolymer-based scaffolds for tissue engineering.

Bio-Based Polymer	Application	Reference
Cydonia oblonga	Glucuronoxylan based hydrogel with a high porous structure, swelling capacity, and biocompatibility as a scaffold for tissue engineering.	[76]
Chitin	Three-dimensional fungal mycelial mats with chitin-glucan polysaccharide cell wall micro-fibrous scaffolds scaffold showed potential application in tissue engineering.	[77]
Chitosan	Chitosan gellan gum hydrogel scaffolds in bone mesenchymal stem cell for tissue engineering application.	[78]
Alginate, cellulose nanocrystals	Variable cross-linking density hydrogel with pH-responsive and reversible network scaffold showed high viability and proliferation abilities with potential application in tissue engineering.	[79]
Chitosan	A composite fiber composed of chitosan and hyaluronic acid by electrospinning and subsequently coated with scaffold showed improvement in cell proliferation.	[80]
Alginate	Ionic cross-linked methacrylated alginate hydrogel microfibers scaffolds with functional 3D tissue structure demonstrated potential growth conditions for cells.	[81]

(Table 2) cont.....

Bio-Based Polymer	Application	Reference
Chitosan	Injectable conductive hydrogel scaffold synthesized using chitosan and κ-carrageenan, with gold nanoparticles enhancing the cell growth and adhesion.	[82]
Chitosan	Injectable *in situ* self-cross linked chitosan-alginic hydrogels demonstrated promising application in cell therapy scaffolds for tissue engineering.	[83]
Anacardium occidentale	Oxidized cashew gum blend with gelatin-based covalently cross-linked hydrogel showed efficacy towards adhesion and proliferation of cells.	[84]
Chitosan	Scaffold for soft tissue engineering fabricated using chitosan and xanthan gum blend showed improved mechanical property and porosity.	[85]
Chitosan	Chitosan, fish scale gelation, and polycaprolactone blend solutions electrospinning scaffold for tissue engineering supported a higher cell adhesion.	[86]
Chitosan	Chitosan-alginate scaffold demonstrated suitability as an alternative substitute for tissue engineering.	[87]
Chitosan	Hydroxyethylcellulose chitosan scaffold showed promising potential for bone tissue engineering application.	[88]
Alginate	Novel bacterial cellulose-alginate scaffold fabricated by lyophilization displayed highly interconnected porous structure with effective cell proliferation.	[89]
Xanthan	Hybrid xanthan and polypyrrole scaffolds with the hydrophobic and elastic network then native xanthan demonstrated cell proliferation.	[90]

Biodegradable Polymeric Composites

A polymeric composite is a multi-phase material consisting of fillers and a polymer matrix that produces suitable structure and mechanical strength for biomedical and industrial applications [72]. Moreover, polymeric matrix composites are also known as reinforced arrangements embedded in the polymer matrix. The hybrid composite materials are generally formed by distributing discontinuous phases within the continuous phase. The continuous phase is known as matrix whereas the discontinuous phase is reinforcement or reinforcing materials. The selection of polymeric matrix materials depends on the nature of the matrix shape, dispersion of reinforcing materials, the interfacial tension between matrix and fillers, and the quality of interface with the production process used [73]. Polymeric composites are classified by the source of the matrix formerly used and the types of the polymer matrix. Further, the source of polymeric composites is sub-classified as a synthetic fiber composite, natural-fiber composite, and bio-composite. While thermoplastic, thermoset, rubber, and bio-based polymers are examples of the types of polymer matrix [74].

Fiber-reinforced is high-performing composites made by cross-linking cellulosic fiber molecules with resins that could be recycled 20 times [75]. The reinforced fiber derived from renewable and carbon dioxide neutral resources is the basic material used for the fabrication of biodegradable natural fiber composites (Fig. **3**). Bio-composites that often mimic the structure of the living materials are formed by combining matrix-resin and reinforced natural fibers. These bio-composites are classified into wood and non-wood fibers.

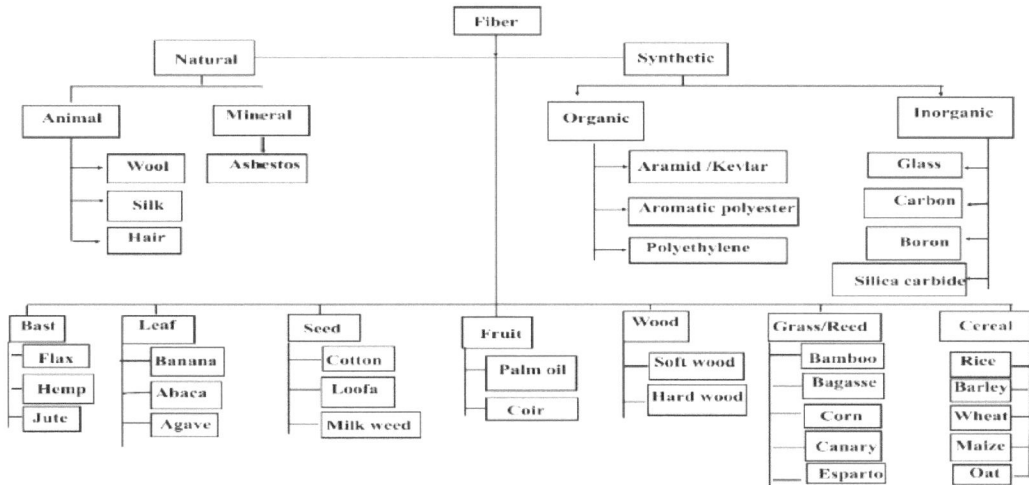

Fig. (3). Classification of fiber source.

The thermoplastic-based composites are fabricated using synthetic polymers. Moreover, thermoplastic polymers produce a thermos-responsive composite that softens at higher temperatures and hardens at lower temperatures. The thermoset composites are commonly fabricated using glass carbon or aramid fibers, incorporated with resins such as polyesters, vinyl esters, epoxides, bismaleimides, cyanate esters, and polyimides or phenolic compounds [91]. While rubber composites are the combination of a rubber matrix and reinforcing materials to produce high strength with flexibility [92].

The development of composites using synthetic polymers had rose a significant concern about the environment for the safety of nature and living beings. Manufacturing industries have drawn considerable attention towards plant-derived bio-based polymers and fibers to replace synthetic material-based composites. Bio-based polymers and especially fibers have been used in the field of textile for many years. However, the application in the fabrication of composites and scaffolds gained attention over the last few decades for their relatively low-cost and environmental friendly nature, compared with synthetic polymers. Recent research reports on the fabrication of a biodegradable composite using bio-based polymers are presented in Table **3**.

Table 3. Composites fabricated using biodegradable polymers with potential biomedical and food preservative applications.

Bio-Based Polymer	Application	Reference
Persian, gellan gums	Sodium caseinate reinforced biodegradable composite showed mechanical and good barrier properties against moisture and light and could have potential application in packaging.	[100]
Chitosan, alginate	Composite hydrogel demonstrated excellent cytocompatibility, blood compatibility, and histocompatibility with the anti-inflammatory response.	[101]
Diverse polysaccharides	Polyphenol incorporated biodegradable composite fabricated using various polysaccharides showed good antioxidant potential with the potential to stop SARS-CoV-2 transmission through the food supply chain.	[102]
Potato starch	Starch-clay blend composite with a range of plasticizers showed improved mechanical strength and reduction in swelling index.	[103]
Corn starch	Starch-chitosan composite showed that a higher composition of starch could significantly produce the film with higher mechanical strength and barrier strength.	[104]
Bacterial cellulose	Bacterial cellulose disintegrated with montmorillonite showed stable water vapor permeability at lower concentrations, however the permeability significantly increased with the increase in the concentration of montmorillonite.	[105]
Pectin	Valorization of pectin from industrial citrus waste was upgraded as biocomposites for mulching application.	[106]
Chitosan	Chitosan-polyvinyl alcohol-containing biogenic metal nanoparticles demonstrated enhancement of food shelf-life.	[107]
Chitosan	Chitosan-polyvinyl alcohol fortified with natural anthocyanins composite showed a good change in color during beverage freshens observation.	[108]
Maize starch	Composite of crystalline starch with agar indicated improvement in mechanical strength and physicochemical properties on the increase of agar content.	[109]
Sugar beet pulp	Pure amylose base transparent composite was fabricated using glycerol as a plasticizer at varied ratios, which could act as an alternative to synthetic bioplastic to reduce plastic waste.	[110]
Alginate, graft chitosan	Multifunctional alginate/carboxymethyl chitosan composite indicated accelerated hemostasis with inhibition in bacterial infections.	[111]
***Morus nigra*, alginate**	Mulberry pekmez-alginate composite showed potential for replacement of non-biodegradable plastic packaging in food applications.	[112]

(Table 3) cont.....

Bio-Based Polymer	Application	Reference
Maize starch	Electrospun composite fiber mats showed improved compatibility among biopolymers and mechanical strength that could potentially be useful in various biomedical applications.	[113]
Pectin	Pectin nanofiber mats cross-linked with adipic acid dihydrazide showed control processing parameters that could effectively help to develop electrospun mats for various biomedical applications.	[114]
Chitosan	Chitosan-kefiran blend nanofiber mats showed excellent biocompatibility and cell supporting functions could be a promising scaffolds for tissue engineering application.	[115]
Psyllium husk	Psyllium husk gelatin-polyvinyl alcohol blend based nanofibers demonstrated efficacy towards potential application for wound dressings and other biomedical applications.	[116]
Alginate	Alginate based electrospun nanofibers showed efficiency in healing and bone reparation.	[117]
Chitosan	Biodegradable chitosan-curdlan electrospun fibers indicated potential application in bone tissue engineering.	[118]
Gellan	Cinnamaldehyde loaded gellan-polyvinyl alcohol electrospun fibers showed promising materials for wound dressings in eradication biofilm.	[119]
Pullulan	Pea protein isolate and pullulan cross-linked electrospun nanofiber films exhibited better hydrophobic properties compared to un-cross linked.	[120]
Pullulan	Pullulan/ethylcellulose composite nanofiber films showed enhanced mechanical strength.	[121]
Amorphophallus konjac	Konjac-polyacrilonitrile nanofiber mats demonstrated absorbent efficacy towards the removal of pollutants in wastewater and as an air purifier for filter media.	[122]
Carboxymethyl cellulose, sodium alginate	Polysaccharides-based polymers such as hyaluronic acid, carboxymethyl cellulose, and sodium alginate electrospun nanofiber mats developed as novel surgical adhesion barriers indicated bio-compatibility with non-adherent cell features.	[123]
Chitosan	Electrospun membranes of chitosan-gelatin blend exhibited a high porosity and surface area.	[124]
Potato protein	Novel potato proteins polysaccharides co-blended with polyethylene oxide electrospun composite fiber indicated a diverse range of applications as advanced functional materials in food science and biomedical fields.	[125]
Chitosan	A bilayer nonwoven fibers for tissue regeneration using chitosan and hyaluronic acid was prepared by electrospinning, which showed good compatibility with mesenchymal stem cells.	[126]
Xanthan, chitosan	Xanthan-chitosan electrospun nanofibers encapsulated with curcumin showed transport through transepithelial and permeation across cell monolayers.	[127]

(Table 3) cont.....

Bio-Based Polymer	Application	Reference
Chitosan	Chitosan intelligent film fortified with phenolic rich rice berry extract for monitoring seafood freshness.	[128]
Chitosan	Chitosan active film fortified with spices extract extended the shelf life of refrigerated pork.	[129]
Chitosan	Chitosan film with hydroxycinnamic acids extended the shelf life of refrigerated pork for 10 additional days.	[130]

Biopolymers in Electrospinning Technology

Several techniques have been developed over the year to fabricate composites including solvent casting, molding, phase separation, centrifugal spinning, and electrospinning. The electrospinning process comprises the application of a strong electrostatic field to a syringe needle linked with a reservoir containing polymeric solution [93]. The syringe needle is connected with an electrode (voltage of 20 kV) that sprays the polymer solution (~0.015 ml/min) over a rotating stainless steel rod kept at a perpendicular distance to form a surface forming Taylor [94]. This polymer solution later converts into the continuous uniform fiber of 10 nm to 10 μm under optimum conditions and is collected over a surface. The fabrication process of fibers *via* electrospinning is affected by various factors such as the polymer molecular weight, concertation, viscosity, conductivity, surface tension, solvent volatility, dielectric constant, applied voltage, and distance between the jet and a collector [95]. Both synthetic and natural polymers alone or in combination have been tested for the fabrication of composites, fibers, and nanofibers with high adaptive, morphology, and function. Due to these multifarious properties of the composite and scaffold developed using electrospinning, they can be employed in several biomedical applications such as wound healing dressing, tissues engineering, *etc.* [96]. However, due to biocompatibility issues, synthetic polymers are less preferred compared to those bio-based polymers. Owing to compatibility, structural, and functional properties of bio-based polymers from the plant as well as from animal sources, these have been exploited for preparing fiber that finds suitability in various biomedical applications. Although bio-based polymers have shown potential in the fabrication of nanofibers using electrospinning, they cannot produce products with excellent mechanical strength. The mechanical properties of bio-based excipients can be enhanced either by crosslinking or blending with other polymers [97]. In addition, altering the physical properties of polymeric electrospinning spray solution such as pH, temperature, *etc.*, further assists in obtaining electrospun nanofibers. Moreover, the selection of the optimum solvent for the preparation of a polymer solution is also considered an important parameter [97]. Generally, both organic, and aqueous solvents are used in the preparation of polymeric solutions including deionized water, poly(hydroxybutylate-cohydroxyvalerate), hexafluoroiso-propa-noal, trifluoro ethanol, dicholoromethane, ethyl acetate, formic acid, *etc*. Electrospinning scaffolds have been incorporated in the sustained delivery system for the release of active pharmaceuticals with improved efficacy. Curcumin incorporated within silk fibroin nanofiber was reported to demonstrate excellent antioxidant and anti-inflammatory actions due to the covalent interaction among each other [98]. Moreover, a different study indicated an acceleration of wound closure rate with silk fibroin incorporated with silver sulfadiazine [99].

Non-Invasive Micro-Needles Fabrication Using Biodegradable Bio-Based Polymer

The transdermal drug delivery system has gained significant attention over the past years in delivering a drug that could demonstrate local and systematic therapeutic effects through the skin. Skin is a living tissue that allows restricted absorption of foreign particulate or medicine. Depending on physicochemical properties of drug molecules that come in contact with the dermal layer of the skin, they penetrate the stratum carenum and can subsequently pass through the viable epidermis, dermis, vascular network, and reach systemic circulation. Prolonged delivery of active pharmaceuticals through the dermal layer has the advantage of bypassing the hepatic first-pass metabolism effect. However, effective delivery of the active pharmaceuticals is affected by several factors including the molecular weight of the drug, substance concentration, skin pathological factors (skin hydration, mechanical damage, skin diseases), lipid content distribution pattern, *etc*. [131]. These permeability associated factors have been successfully bypassed through the innovation of painless non-invasive needles. Lateron, the transformation in fabrication strategy of microneedles array patch from synthetic to biodegradable nano-needles has gained attention among researchers and pharmaceutical manufacturers.

Microneedles are devices composed of several micron-sized needles organized in a structural assembly of an array to deliver the pharmaceuticals for localized applications. These tapered sharp tip micron-sized needles are present in a dimension of 50 to 250 μm in width and length that could range from micrometer to 1500 μm in length [132]. Microneedles are generally fabricated using metal, silicon, glass, ceramic, and polymers. Several reports indicated that polymers-based microneedles could deliver the active medicament in a regulated form. However, the demand and the use of a synthetic polymer that could not degrade on the site of application have reduced significantly. In the meantime, bio-based polymers were developed as an alternative to synthetic polymers for the fabrication of micro needle array patches. Biodegradable polymers are a frequently used terminology for "biopolymers" derived from natural resources.

Moreover, bio-based polymers have a major advantage over synthetic polymers in terms of degradation. In addition, biopolymers fabricated in a dosage form can degrade inside the living organ or tissue without any toxic effects.

Both synthetic and natural polymers have demonstrated utility in the fabrication of solid microneedles, coated microneedles, dissolving microneedles, and hollow microneedles. Biodegradable polymers including chitosan [133], starch [134], polyvinyl alcohol [135], polyvinylpyrrolidone [136], alginate [137], carboxymethyl cellulose [136], hyaluronic acid [137], pullulan [138], gelatin [134] and more have been extensively studied by the researcher in the development of microneedles array patch. Recently biodegradable protein-based polymers from fish scales with similar physiochemical properties to gelatin have been discovered and explored in the fabrication of microneedles [139]. In addition, biodegradable polymers obtained from fish scales have also been explored for the development of eco-friendly plastic. Das *et al.* extracted a biodegradable polymer from the fish scales of *Tilapia* and fabricated patient-friendly lidocaine incorporated microneedles as a substitute for a conventional transdermal dosage form [140]. Diffusion study through porcine skin indicated a 2.5 to 7.5% increase in permeation of drug through microneedles [140]. Another study on nano-cellulose isolated from the *Tilapia* scale demonstrated the formation of excellent microneedles *via* high-temperature micro-molding techniques with regulated dissolution needles for prolonging delivery of pharmaceuticals [141]. In addition, the mechanical property of film produced using fish scale biopolymer showed Young's modulus 0.23 N/mm^2 and a tensile strength of 1.8105 N/mm^2 [142]. Moreover, microneedles were fabricated using fish scales incorporated with a lipophilic drug that showed enhanced permeability of drug through pork skin [143]. Recent reports on the bio-based polymer isolated from natural resources in the fabrication of microneedles with its use in drug delivery systems have been presented in Table **4**.

Table 4. Application of bio-based polymers isolated from natural resources in the fabrication of microneedles.

Bio-Based Polymer	Application	Reference
Fish scale	Cross-linked hydrolyzed collagen loaded with ferrous gluconate showed excellent strength before insertion in the skin with fast dissolution after insertion.	[144]
Fish scale	Hydrolyzed collagen-based transdermal film loaded with aspirin demonstrated enhancement in dermal permission of aspirin.	[143]
Fish scale	Lidocaine loaded biodegradable microneedles fabricated using hydrolyzed collagen indicated a slow permeation through the skin.	[140]

(Table 4) cont.....

Bio-Based Polymer	Application	Reference
Fish scale	Microneedles produced using nano-cellulose loaded with fish scale biopolymer demonstrated lower elongation with higher tensile strength compared to native biopolymer.	[141]
Fish scale	Microneedles prepared using fish scale biopolymer indicated that needles can effectively pierce and degrade into the skin.	[142]
Polyvinyl alcohol	Polyvinyl alcohol microneedle capped with sodium bicarbonate and polyvinyl pyrrolidone demonstrated separation of the needle within 90 s with a promising platform for transdermal delivery.	[145]
Pullulan	Pullulan based dissolving microneedles showed potential towards efficient delivery of low molecular weight across the skin.	[146]
Polyvinyl pyrrolidone	Polyvinyl pyrrolidone based biodegradable microneedle arrays loaded with high molecular weight polypeptide and α-choriogonadotropin, were fabricated for systemic delivery.	[147]
Silkworm cocoons of ***Bombyx mori***	Silk fibroin microneedle patches loaded with insulin demonstrated sustained release and could be effective in the treatment of diabetes.	[148]
Panax notoginseng	Microneedles fabricated using natural derived immunoactive macromolecules showed satisfactory mechanical strength with required skin penetration depth.	[149]
Starch and gelatin	Starch/gelatin microneedles demonstrated rapid and efficient delivery of insulin in the management of diabetes.	[134]
Bletilla striata	Microneedles fabricated using natural polysaccharides were more efficacious in transdermal delivery of the drug compared with conventional patches.	[150]
Carboxymethyl cellulose	Carbohydrate polymer fabricated microneedles demonstrated biocompatibility with the biodegradability.	[151]
Alginate	Calcium ion cross-linked alginate/maltose composite microneedles encapsulated with insulin indicated the effective hypoglycemic effect.	[152]

Green Synthesis of Metallic Nanoparticles Using Bio-Based Excipient

Nanotechnology is one of the most promising technologies of the 21st century. However, it is deemed necessary to differentiate between nanoscience and nanotechnology. Generally, nanoscience deals with handling or manipulation of materials at atomic and molecular scales, whereas nanotechnology is the ability to observe, quantify, manipulate, assemble, control, and manufacture materials at the nanometer scale. Moreover, nanotechnology also refers to a broad science that deals with the control of particles, electrons, protons, and neutrons. The prefix "*nano*" referred to the Greek word denoting "*dwarf*" or a small size of materials that represents one thousand million parts of a meter (10^{-9} m) [153]. Nanotechnologies contribute to almost every field of science, including physics, material science, chemistry, biology, computer science, and engineering. Progress starting from the ancient era in the field of nanotechnology has been presented in Fig. **(4)**. The history of metallic nanoparticles is very old and the use of nanoparticles as coloring agents was discovered in the ancient Roman era during the fourth century [154]. This color produced by metallic nanoparticles was due to Surface Plasmon Resonance and optical properties, also known as specific property that does not change or fade with time. Faraday in 1857 later recognized the existence of metallic nanoparticles in a solution and the quantitative explanation on color was proven by Mie in 1908 [155].

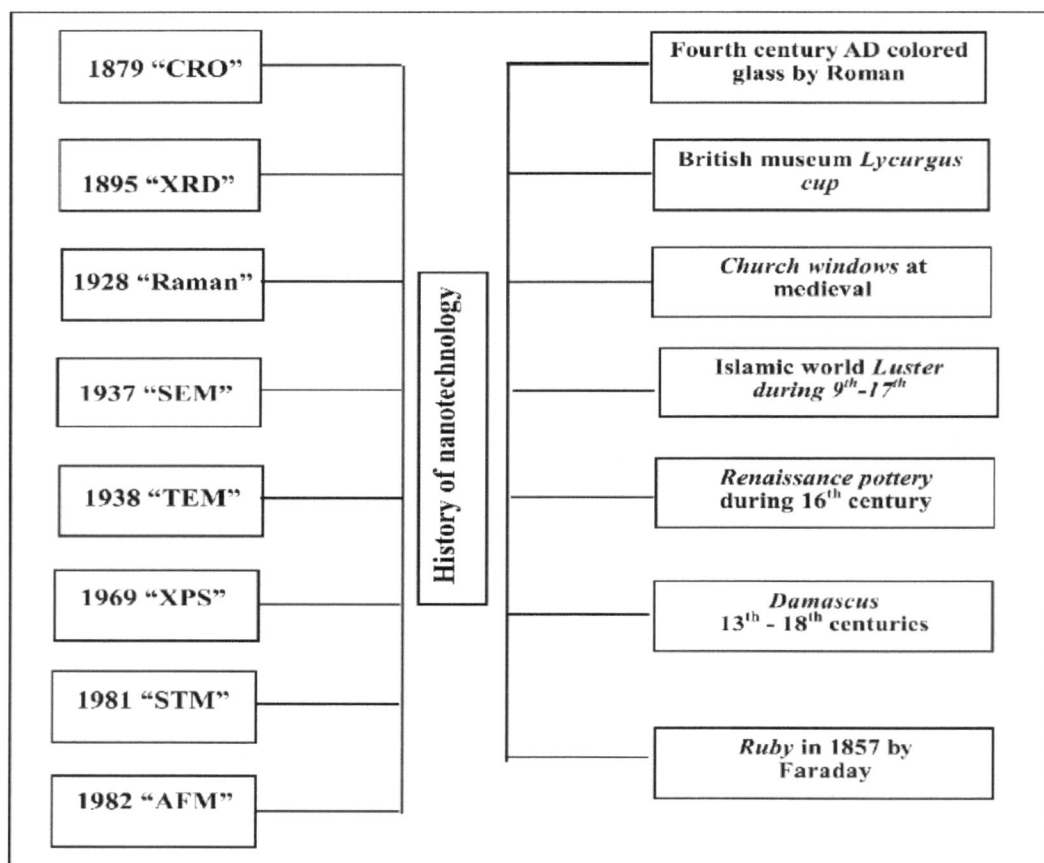

Fig. (4). Progress in nanotechnology.

Tremendous results of metallic nanoparticles especially in the treatment of cancer have attracted researchers to develop biocompatible nanomaterials. Nanomaterials for biomedical applications are fabricated using either synthetic organic or inorganic chemicals or *via* green chemistry. However, concerning the biocompatibility, biodegradability, and accessibility of plant extracts and bio-based polymers, they have gained significance compared to chemicals in the reduction of fabrication of nanomaterials. The rise of antibiotic resistance in various microorganisms demands the development of multifarious biopolymers-based nanoparticles. Thus, possibly antibacterial efficacy of therapeutics can be synergized with simultaneous incorporation of metallic nanoparticles within novel drug delivery systems.

Cellulose extracted from a natural resource such as plants and other living organisms has been utilized as a capping and reducing metallic material to nanoparticles [156]. Spagnol and coworkers modified the surface structure of cellulose Nano whisker using succinic anhydride and converted it in the form of nanofibers coated biogenic silver nanoparticles, demonstrating excellent antibacterial efficacy [157]. Hajji and coworkers fabricated a dressing with chitosan and polyvinyl alcohol *in situ* to reduce and stabilize silver nitrate to nanoparticles, which indicated improved mechanical strength and antibacterial efficacy against potential wound pathogens. Recent progress in the fabrication of metallic nanomaterials using bio-based polymer is presented in Table **5**.

Table 5. Recently reported green synthesized metallic nanoparticles with the potential antibacterial application.

Bio-Based Polymer	Synthesis of Nanoparticle and Application	Reference
Ramaria botrytis	Biogenic synthesis of bimetallic nanoparticles showed antioxidant and antibacterial efficacy with catalytic activity towards 4-nitrophenols.	[158]
Polysaccharide from macro-algae	Silver nanoparticles synthesized using polysaccharides extracted from macro-algae and applied over cotton fabrics demonstrated process dependent antimicrobial activity.	[159]
Alginate	Seaweed polysaccharide catalyzed silver nanoparticles showed concentration-dependent biocompatibility with considerable antibacterial properties.	[160]
Pectin	Pectin capped silver nanoparticles exhibited significant antimicrobial activity against both gram-negative and positive bacteria.	[161]
Corn cobs	Xylan isolated from corn cob capped silver nanoparticles showed significant anti *Trypanosoma cruzi* activity.	[162]
Exopolysaccharide	Biogenic silver nanoparticles synthesized using exopolysaccharides from *Streptomyces violaceus* MM72 showed enhanced antioxidant and promising antibacterial activity.	[163]
Exopolysaccharide	Biogenic silver nanoparticles synthesized using *Bacillus subtilis* MTCC 2422 exopolysaccharides demonstrated higher adherence towards the bacterial surface with good bactericidal activity.	[164]

(Table 5) cont.....

Bio-Based Polymer	Synthesis of Nanoparticle and Application	Reference
Tamarindus indica	Carboxymethylated tamarind polysaccharide capped silver nanoparticles showed concentration-dependent cytotoxicity against mammalian cells with significant inhibition of growth and biofilm of both gram-negative and positive.	[165]
Sodium alginate	Sodium alginate/polyaniline boronic acid capped gold nanoparticles/nanocomposite demonstrated moderate antioxidant efficacy and biocompatibility with human cell lines.	[166]
Sodium alginate	Sodium alginate stabilized gold nanoparticles demonstrated excellent colorimetric detection of ascorbic acid.	[167]
Sodium alginate	Sodium alginate/poly(3-aminophenyl boronic acid) stabilized silver nanoparticles incorporated with nanogel showed good sensing potential for hydrogen peroxide in water.	[168]
Opuntia ficus-indica	Green mucilage capped silver nanoparticles and modified glassy carbon electrode composite showed efficient sensing of glucose with a fuel cell application.	[169]
Linum usitatissimum	Mucilage/protein stabilized silver nanoparticles with antibacterial properties showed potential toward application health and medical field.	[170]
Abelmoschus esculentus	Zinc oxide nanoparticles synthesized using plant mucilage showed selective photocatalyst for degradation of cationic dye.	[171]

Other Applications

The term polymer sounds very specific to the pharmaceutical industry, however polymers have a wide application in various industrial sectors including aerospace, automotive, electronic, packaging, biosensors, and fabrication of medical devices. Moreover, application of polymers has also shown excellent results in sports equipment, 3D plastic printing, water purification, printed circuit board substrates, renewable biomass operations, insulators, solar panels, medical devices, contact lenses, fabrication of bulletproof, fire-resistant, and lifesaving jackets [172]. In addition, due to the presence of diverse elements within polymers, several other formulation based applications have been reported such as an emulsifier, a suspending agent, a disintegrating agent, a filler, a flow property enhancer, in molecular regeneration, cosmetics, angioplasty, vascular stents, ventricular assist devices, *etc.* [172]. Natural polymers such as proteins, starch, cellulose, and gum are freely available for centuries and have been applied as materials for food, leather, fibers, waterproofing, and coating. The percentage-wise application of bio-based polymers has been presented in Fig. (**5**) as obtained from the Scopus database.

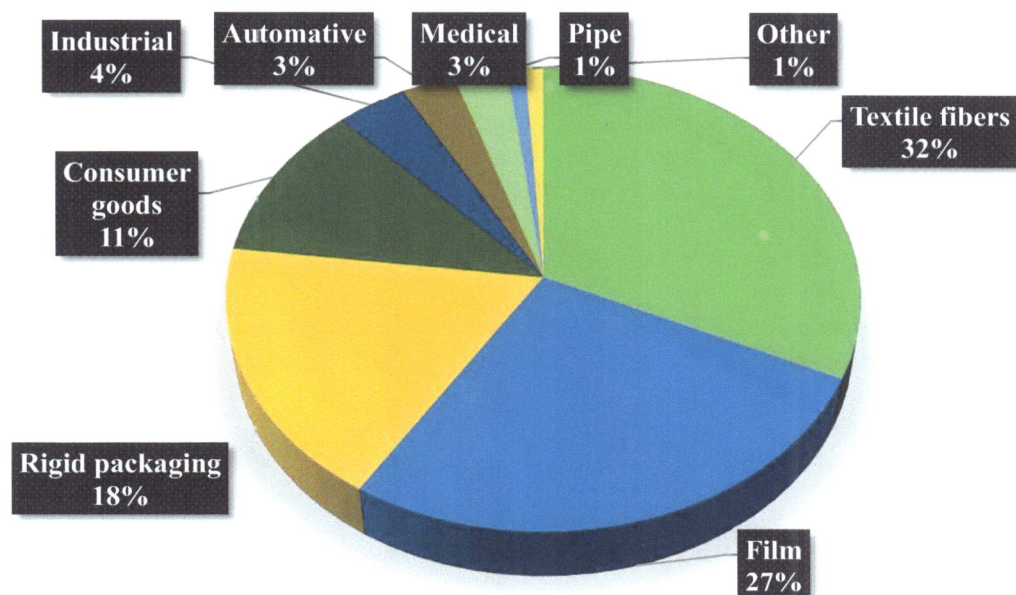

Fig. (5). Application of bio-based polymers in the various field available online on the Scopus database. The search includes Bullen terms such as "polysaccharides*", "bio-based polymer*", "mucilage" (Assessed on 17th Aug 2021) (for interpretation of the results to color in this Fig legend, the reader referred to either web version of this chapter or color print).

CONCLUSIONS AND FUTURE PROSPECTS

Research on bio-based polymers has demonstrated great success over the past few years in safe and effective delivery of pharmaceutical active ingredients to treat a range of medical conditions. Biopolymers are inexhaustible and sustainable natural polymers available freely in nature. Progress in the advancement of biopolymers with their application in various sectors has expanded drastically due to biocompatibility and biodegradability. This progression needs further advancements in polymer science based on the modification of the chemical and physical properties of biopolymers as required by the innovator. The research activities highlighted in this chapter showed that the potential application of bio-polymers is not limited to drug delivery but has also expanded to several relevant fields. Moreover, polymeric materials have a vast potential for stirring new applications in the area of storage and conduction of electricity, heat, and light. In addition, polymers have potential application in molecular-based information storage and processing, molecular composites, separation of the membrane, packaging, housing, and transportation.

The pharmacological efficacy of therapeutics fabricated as nanostructure materials capped with bio-based polymers is widely accepted to treat various diseases.

Tissue or organ targeting *via* targeted delivery integrated nanomaterials indicated excellent therapeutic effects against a wide range of bacterial infections. However, due to rising resistance of antibiotics either in conventional or novel dosage form delivery demands the incorporation of synergistic nanomaterials within the dosage form. Green synthesized metallic nanoparticles demonstrated biocompatibility and broad antimicrobial efficacy against various pathogens. Moreover, metallic materials including silver and gold nanoparticles showed excellent antibacterial efficacy against the last line of antibiotic therapy. The combination of metallic nanoparticles with polymeric incorporated antibiotics or deactivated bacteria could be a potential future for the treatment of bacterial infections and cancer that are becoming resistant day by day to effective antibiotics.

REFERENCES

[1]   Rao MG, Bharathi P, Akila R. A comprehensive review on biopolymers. Sci Revs Chem Commun 2014; 4(2): 61-8.

[2]   Kamal IM. Polymers in concrete. Soran University, Faculty of Engineering, Chemical Engineering Department 2014. Available from: https://www.researchgate.net/profile/Ibtisam_Kamal/publication/281110586_Polymers_In_Concrete/links/55d60f0608ae9d65948bacc5/Polymers-In-Concrete.pdf

[3]   Premraj R, Doble M. Biodegradation of polymers. Indian J Biotechnol 2005; 4: 186-93.

[4]   Yadav P, Yadav H, Shah VG, Shah G, Dhaka G. Biomedical biopolymers, their origin and evolution in biomedical sciences: a systematic review. J Clin Diagn Res 2015; 9(9): ZE21-5.
[http://dx.doi.org/10.7860/JCDR/2015/13907.6565] [PMID: 26501034]

[5]   Hua S. Advances in oral drug delivery for regional targeting in the gastrointestinal tract-Influence of physiological, pathophysiological and pharmaceutical factors. Front Pharmacol 2020; 11: 524.
[http://dx.doi.org/10.3389/fphar.2020.00524] [PMID: 32425781]

[6]   Fasinu P, Pillay V, Ndesendo VMK, du Toit LC, Choonara YE. Diverse approaches for the enhancement of oral drug bioavailability. Biopharm Drug Dispos 2011; 32(4): 185-209.
[http://dx.doi.org/10.1002/bdd.750] [PMID: 21480294]

[7]   Abay F, Ugurlu T. Orally disintegrating tablets: a short review. J Pharm Drug Develop 2015; 3(3): 303.

[8]   Sukhavasi S, Kishore VS. Formulation and evaluation of fast dissolving tablets of amlodipine besylate by using Fenugreek seed mucilage and *Ocimum basilicum* gum. Int Curr Pharm J 2012; 1(9): 243-9.
[http://dx.doi.org/10.3329/icpj.v1i9.11614]

[9]   Bucktowar K, Bucktowar S, Bucktowar M, Bholoa LD, Ganesh N. Formulation and evaluation of fast dissolving tablets of paracetamol using *Ocimum basilicum* seed mucilage as superdisintegrant. Int Res J Pharm Biosci 2017; 4(1): 44-55.

[10]   Zafar N, Neeharika V, Lakshmi PK. Formulation and evaluation of sildenafil citrate fast dissolving tablets using fenugreek seed mucilage. Int J Res Ayurveda Pharm 2014; 5(3): 352-8.
[http://dx.doi.org/10.7897/2277-4343.05373]

[11]   Ravi S, Reddy I. formulation and evaluation of domperidone fast dissolving tablets using *Plantago ovata* mucilage. Int J Pharm Sci Res 2013; 3489.

[12]   Shirsand SB, Suresh S, Para MS, Swamy PV, Kumar DN. *Plantago ovata* mucilage in the design of fast disintegrating tablets. Indian J Pharm Sci 2009; 71(1): 41-5.
[http://dx.doi.org/10.4103/0250-474X.51952] [PMID: 20177454]

[13]   Swathi S, Neeharika V, Lakshmi PK. Formulation and evaluation of fast dissolving tablets of freely

and poorly soluble drug with natural and synthetic super disintegrants. Drug Invent Today 2011; 3(10): 250-6.

[14]   Prasad Panda B, Gregory J. Extraction and performance evaluation of *Salvia hispanica* mucilage as natural disintegrants for optimization of pyrilamine maleate fast dissolving tablets. Nat Prod J 2015; 5(4): 288-98.
[http://dx.doi.org/10.2174/2210315505666150914224314]

[15]   Sukhavasi S, Kishore VS. Formulation and evaluation of fast dissolving tablets of amlodipine besylate by using *Hibiscus rosa-sinensis* mucilage and modified gum karaya. Int J Pharm Sci Res 2012; 3(10): 3975.

[16]   Patil PS, Badgujar SV, Shirsath KG, Sonawane MS. Evaluating the efficacy of *Hibiscus sabdariffa* linn mucilage as binder in orodispersible tablets of losartan potassium. J Pharm Advanc Res 2021; 4(3): 1173-8.

[17]   Sharma G, Bhurat M, Shastry V, Shrivastava B. Design and development of metoprolol succinate sustained release tablet using *Eulophia herbacae* and *Remusatia vivipara* tuber mucilage as novel drug release modifiers. Mater Technol 2021; 1-11.
[http://dx.doi.org/10.1080/10667857.2021.1967550]

[18]   Narkhede Sachin B, Vidyasagar G, Jadhav Anil G, Bendale Atul R, Patel Kalpen N. Isolation and evaluation of mucilage of *Artocarpus heterophyllus* as a tablet binder. J Chem 2010; 2(6): 161-6.

[19]   Ramu G, Mohan GK, Jayaveera KN. Preliminary investigation of patchaippasali mucilage (*Basella alba*) as tablet binder. International Journal of Green Pharmacy 2011; 5(1): 24.
[http://dx.doi.org/10.4103/0973-8258.82091]

[20]   Olayemi O, Jacob O. Preliminary evaluation of *Brachystegia eurycoma* seed mucilage as tablet binder. Int J Pharm Res Innov 2011; 3(1): 01-6.

[21]   Kale R, Joshi U, Ambhore D, Sitaphale G. Evaluation of *Delonix regia* Raf. endospermic mucilage as tablet binder. Int J Chemtech Res 2009; 1(1): 11-5.

[22]   Ahuja M, Kumar A, Yadav P, Singh K. *Mimosa pudica* seed mucilage: Isolation; characterization and evaluation as tablet disintegrant and binder. Int J Biol Macromol 2013; 57: 105-10.
[http://dx.doi.org/10.1016/j.ijbiomac.2013.03.004] [PMID: 23500434]

[23]   Rahim H, Sadiq A, Khan S, *et al.* Isolation and preliminary evaluation of *Mulva Neglecta* mucilage: a novel tablet binder. Braz J Pharm Sci 2016; 52(1): 201-10.
[http://dx.doi.org/10.1590/S1984-82502016000100022]

[24]   Venditti I. Morphologies and functionalities of polymeric nanocarriers as chemical tools for drug delivery: A review. J King Saud Univ Sci 2019; 31(3): 398-411.
[http://dx.doi.org/10.1016/j.jksus.2017.10.004]

[25]   Ishak RAH. Buoyancy-generating agents for stomach-specific drug delivery: an overview with special emphasis on floating behavior. J Pharm Pharm Sci 2015; 18(1): 77-100.
[http://dx.doi.org/10.18433/J3602K] [PMID: 25877444]

[26]   Hanif S, Sarfraz RM, Syed MA, Ali S, Iqbal Z, Shakir R, *et al.* Formulation and Evaluation of chitosan-based polymeric biodegradable mucoadhesive buccal delivery for locally acting drugs: *in vitro, ex vivo* and *in vivo* volunteers characterization. Lat Am J Pharm 2021; 40(4): 670-81.

[27]   Acartürk F. Mucoadhesive vaginal drug delivery systems. Recent Pat Drug Deliv Formul 2009; 3(3): 193-205.
[http://dx.doi.org/10.2174/187221109789105658] [PMID: 19925443]

[28]   Hanafy NAN, Leporatti S, El-Kemary M. Mucoadhesive curcumin crosslinked carboxy methyl cellulose might increase inhibitory efficiency for liver cancer treatment. Mater Sci Eng C 2020; 116: 111119.
[http://dx.doi.org/10.1016/j.msec.2020.111119] [PMID: 32806233]

[29]   Abdul Ahad H, Chinthaginjala H, Priyanka MS, Raghav DR, Gowthami M, Jyothi VN. *Datura stramonium* leaves mucilage aided buccoadhesive films of aceclofenac using 3^2 factorial design with design-expert software. Indian Journal of Pharmaceutical Education and Research 2021; 55(2s): s396-404.
[http://dx.doi.org/10.5530/ijper.55.2s.111]

[30]   Pardeshi CV, Dhangar RN, Jagatap VR, Sonawane RO. New cationic neuronanoemulsion-laden *Eulophia herbacea* mucilage based mucoadhesive hydrogel for intranasal delivery of chlorpromazine. Mater Technol 2021; 36(4): 189-202.
[http://dx.doi.org/10.1080/10667857.2020.1740859]

[31]   Singh S, Dodiya TR, Dodiya R, Singh S, Bothara SB. *In vivo, ex vivo*, and *in vitro* mucoadhesive strength assessment of potential pharmaceutical bio-resource polymer from *Diospyros melonoxylon* roxb seeds. International Journal of Pharmaceutical Sciences and Nanotechnology 2021; 14(1): 5307-14.
[http://dx.doi.org/10.37285/ijpsn.2021.14.1.6]

[32]   Ray P, Chatterjee S, Saha P. Screening of polysaccharides from fruit pulp of *Ziziphus mauritiana* L. and *Artocarpus heterophyllus* L. as natural mucoadhesives. Future Journal of Pharmaceutical Sciences 2021; 7(1): 29.
[http://dx.doi.org/10.1186/s43094-020-00164-5]

[33]   Singh S, Sunil BB, Ayaz A. Formulation of oral mucoadhesive tablets using mucilage isolated from *Buchanania lanzan* spreng seeds. Int J Pharm Sci Nanotech 2014; 7(2): 2494-503.
[http://dx.doi.org/10.37285/ijpsn.2014.7.2.11]

[34]   Singh S, Sunil BB. Development and evaluation of oral mucoadhesive tablets of losartan potassium using natural gum from *Manilkara zapota*. Int J Pharm Sci Nanotech 2013; 6(4): 2245-54.
[http://dx.doi.org/10.37285/ijpsn.2013.6.4.7]

[35]   Irgang M, Sauer IM, Karlas A, *et al*. Porcine endogenous retroviruses: no infection in patients treated with a bioreactor based on porcine liver cells. J Clin Virol 2003; 28(2): 141-54.
[http://dx.doi.org/10.1016/S1386-6532(02)00275-5] [PMID: 12957184]

[36]   Iravani S, Varma RS. Plants and plant-based polymers as scaffolds for tissue engineering. Green Chem 2019; 21(18): 4839-67.
[http://dx.doi.org/10.1039/C9GC02391G]

[37]   Tavares-Negrete JA, Aceves-Colin AE, Rivera-Flores DC, *et al*. Three-dimensional printing using a maize protein: zein-based inks in biomedical applications. ACS Biomater Sci Eng 2021; 7(8): 3964-79.
[http://dx.doi.org/10.1021/acsbiomaterials.1c00544] [PMID: 34197076]

[38]   Azimi B, Maleki H, Zavagna L, *et al*. Bio-based electrospun fibers for wound healing. J Funct Biomater 2020; 11(3): 67.
[http://dx.doi.org/10.3390/jfb11030067] [PMID: 32971968]

[39]   Sanaei Moghaddam Sabzevar Z, Mehrshad M, Naimipour M. A biological magnetic nano-hydrogel based on basil seed mucilage: study of swelling ratio and drug delivery. Iran Polym J 2021; 30(5): 485-93.
[http://dx.doi.org/10.1007/s13726-021-00905-0]

[40]   Bakr RO, Amer RI, Attia D, *et al*. *In-vivo* wound healing activity of a novel composite sponge loaded with mucilage and lipoidal matter of Hibiscus species. Biomed Pharmacother 2021; 135: 111225.
[http://dx.doi.org/10.1016/j.biopha.2021.111225] [PMID: 33434856]

[41]   Kumar V, Mazumder B, Sharma PP, Ahmed Y. Pharmacokinetics and hypoglycemic effect of gliclazide loaded in Isabgol husk mucilage microparticles. J Pharm Investig 2021; 51(2): 159-71.
[http://dx.doi.org/10.1007/s40005-020-00494-9]

[42]   Mohseni F, Goli A, Abdollahi M. Characterization of tertiary conjugate of gelatin-flaxseed (*Linum*

usitatissimum) mucilage- oxidized tannic acid. Food Sci Technol (Campinas) 2021; 17(108): 59-73.
[http://dx.doi.org/10 52547/fsct.17.108.59]

[43] Sharma VK, Sharma PP, Mazumder B, *et al.* Mucoadhesive microspheres of glutaraldehyde crosslinked mucilage of Isabgol husk for sustained release of gliclazide. J Biomater Sci Polym Ed 2021; 32(11): 1420-49.
[http://dx.doi.org/10.1080/09205063.2021.1925389] [PMID: 33941041]

[44] Mahor S, Chandra P, Prasad N. Neelkant Prasad. Design and *in-vitro* evaluation of float-adhesive famotidine microspheres by using natural polymers for gastroretentive properties. Indian Journal of Pharmaceutical Education and Research 2021; 55(2): 407-17.
[http://dx.doi.org/10.5530/ijper.55.2.78]

[45] Saidin NM, Anuar NK, Tin Wui W, Meor Mohd Affandi MMR, Wan Engah WR. Skin barrier modulation by *Hibiscus rosa-sinensis* L. mucilage for transdermal drug delivery. Polym Bull 2021.

[46] Allafchian AR, Kalani S, Golkar P, Mohammadi H, Jalali SAH. A comprehensive study on *PLANTAGO OVATA*/PVA biocompatible nanofibers: Fabrication, characterization, and biological assessment. J Appl Polym Sci 2020; 137(47): 49560.
[http://dx.doi.org/10.1002/app.49560]

[47] Hosseini MS, Nabid MR. Synthesis of chemically cross-linked hydrogel films based on basil seed (*Ocimum basilicum* L.) mucilage for wound dressing drug delivery applications. Int J Biol Macromol 2020; 163: 336-47.
[http://dx.doi.org/10.1016/j.ijbiomac.2020.06.252] [PMID: 32615215]

[48] Yari K, Akbari I, Yazdi SAV. Development and evaluation of sodium alginate-basil seeds mucilage beads as a suitable carrier for controlled release of metformin. Int J Biol Macromol 2020; 159: 1-10.
[http://dx.doi.org/10.1016/j.ijbiomac.2020.04.111] [PMID: 32330501]

[49] Bal T, Swain S. Microwave assisted synthesis of polyacrylamide grafted polymeric blend of fenugreek seed mucilage-Polyvinyl alcohol (FSM-PVA-g-PAM) and its characterizations as tissue engineered scaffold and as a drug delivery device. Daru 2020; 28(1): 33-44.
[http://dx.doi.org/10.1007/s40199-019-00237-8] [PMID: 30712231]

[50] Bhardwaj S, Bhatia S. Development and Characterization of niosomal gel system using *Lallementia royaleana* Benth. mucilage for the treatment of rheumatoid arthritis. Iran J Pharm Res 2020; 19(3): 465-82.
[PMID: 33680045]

[51] Ghumman SA, Bashir S, Noreen S, Khan AM, Malik MZ. Taro-corms mucilage-alginate microspheres for the sustained release of pregabalin: *In vitro* & *in vivo* evaluation. Int J Biol Macromol 2019; 139: 1191-202.
[http://dx.doi.org/10.1016/j.ijbiomac.2019.08.100] [PMID: 31415852]

[52] Medina-Torres L, Núñez-Ramírez DM, Calderas F, *et al.* Curcumin encapsulation by spray drying using *Aloe vera* mucilage as encapsulating agent. J Food Process Eng 2019; 42(2): e12972.
[http://dx.doi.org/10.1111/jfpe.12972]

[53] Maru SGSS. Physicochemical and mucoadhesive strength characterization of natural polymer obtained from leaves of *Aloe vera.* Pharmtechmedica 2013; 2(3): 303-8.

[54] Hasnain MS, Rishishwar P, Rishishwar S, Ali S, Nayak AK. Extraction and characterization of cashew tree (Anacardium occidentale) gum; use in aceclofenac dental pastes. Int J Biol Macromol 2018; 116: 1074-81.
[http://dx.doi.org/10.1016/j.ijbiomac.2018.05.133] [PMID: 29791876]

[55] Singh S, Bothara SB. Formulation development of oral mucoadhesive tablets of losartan potassium using mucilage isolated from *Diospyros melonoxylon* Roxb Seeds. International Journal of Pharmaceutical Sciences and Nanotechnology 2013; 6(3): 2154-63.
[http://dx.doi.org/10.37285/ijpsn.2013.6.3.7]

[56]   Singh S, Tanvi RD, Sangeeta S, Rajesh D, Sunil BB. Bio-based polymer isolated from seeds of *Buchanania lanzan* spreng with potential use as pharmaceutical mucoadhesive excipient. Int J Pharm Sci Nanotech 2020; 13(4): 5034-41.
[http://dx.doi.org/10.37285/ijpsn.2020.13.4.9]

[57]   Tyagi N. Formulation and evaluation of zidovudine nanosuspensions using a novel bio polymer from the seeds of *Buchanania lanzan.* J Drug Deliv Ther 2013; 3(4): 85-8.
[http://dx.doi.org/10.22270/jddt.v3i4.544]

[58]   Singh S, Nwabor OF, Ontong JC, Voravuthikunchai SP. Characterization and assessment of compression and compactibility of novel spray-dried, co-processed bio-based polymer. J Drug Deliv Sci Technol 2020; 56: 101526.
[http://dx.doi.org/10.1016/j.jddst.2020.101526]

[59]   Singh S, Santoki R. Development and appraisal of mucoadhesive tablets of hydralazine using isolated mucilage of *Annona squamosa* seeds. Int J Pharm Sci Nanotech 2016; 9(4): 3349-56.

[60]   Singh S, Goswami H. Formulation of oral mucoadhesive tablets of pioglitazone using natural gum from seeds of *Pithecellobium dulce.* International Journal of Pharmaceutical Sciences and Nanotechnology 2015; 8(4): 3031-8.
[http://dx.doi.org/10.37285/ijpsn.2015.8.4.6]

[61]   Singh S, Bothara SB. Development of oral mucoadhesive tablets of losartan potassium using natural gum from Manilkara zapota seeds: A potential source of natural gum.ISRN Pharmaceutics 2014; 1-10.

[62]   Singh S, Bothara SB. Development of oral mucoadhesive tablets of losartan potassium using natural gum from *Manilkara zapota* seeds. International Journal of Pharmaceutical Sciences and Nanotechnology 2013; 6(4): 2245-54.
[http://dx.doi.org/10.37285/ijpsn.2013.6.4.7]

[63]   Bothara SB, Singh S. Pharmacognostic studies of seds on some plants belonging chhattisgarh. Pharmacogn J 2012; 4(28): 24-30.
[http://dx.doi.org/10.5530/pj.2012.28.5]

[64]   Singh S, Nwabor OF, Ontong JC, Kaewnopparat N, Voravuthikunchai SP. Characterization of a novel, co-processed bio-based polymer, and its effect on mucoadhesive strength. Int J Biol Macromol 2020; 145: 865-75.
[http://dx.doi.org/10.1016/j.ijbiomac.2019.11.198] [PMID: 31783076]

[65]   Ghumman SA, Bashir S, Noreen S, Khan AM, Riffat S, Abbas M. Polymeric microspheres of okra mucilage and alginate for the controlled release of oxcarbazepine: *In vitro* & *in vivo* evaluation. Int J Biol Macromol 2018; 111: 1156-65.
[http://dx.doi.org/10.1016/j.ijbiomac.2018.01.058] [PMID: 29337102]

[66]   Navamanisubramanian R, Nerella R, D C, Seetharaman S. Shanmuganathan Seetharaman. Use of okra mucilage and chitosan acetate in verapamil hydrochloride buccal patches development: *in vitro* and *ex vivo* characterization. J Young Pharm 2017; 9(1): 94-9.
[http://dx.doi.org/10.5530/jyp.2017.9.18]

[67]   Ghumman SA, Bashir S, Ahmad J, Hameed H, Khan IU. *Colocasia esculenta* corms mucilage-alginate microspheres of oxcarbazepine: design, optimization and evaluation. Acta Pol Pharm 2017; 74(2): 505-17.
[PMID: 29624256]

[68]   Singh S, Sangeet S, Bothara SB, Roshan P. Pharmaceutical characterization of some natural excipient as potential mucoadhesives agent. T Pharm Res 2010; 4: 91-104.

[69]   Singh S, Maru SG, Bothara B S. Formulation and *ex-vivo* evaluation of glipizide buccal tablets. International Journal of Pharmaceutical Sciences and Nanotechnology 2014; 7(2): 2487-93.
[http://dx.doi.org/10.37285/ijpsn.2014.7.2.10]

[70]   Langer R, Vacanti JP. Tissue Engineering. Science 1993; 260(5110): 920-6.

[http://dx.doi.org/10.1126/science.8493529] [PMID: 8493529]

[71] Gentile P, Chiono V, Carmagnola I, Hatton P. An overview of poly(lactic-co-glycolic) acid (PLGA)-based biomaterials for bone tissue engineering. Int J Mol Sci 2014; 15(3): 3640-59.
[http://dx.doi.org/10.3390/ijms15033640] [PMID: 24590126]

[72] Ilyas RA, Sapuan SM. The preparation methods and processing of natural fibre bio-polymer composites. Curr Org Synth 2020; 16(8): 1068-70.
[http://dx.doi.org/10.2174/1570179416082001201005616] [PMID: 31984916]

[73] Bledzki A, Gassan J. Composites reinforced with cellulose based fibres. Prog Polym Sci 1999; 24(2): 221-74.
[http://dx.doi.org/10.1016/S0079-6700(98)00018-5]

[74] Saba N, Tahir P, Jawaid M. A review on potentiality of nano filler/natural fiber filled polymer hybrid composites. Polymers (Basel) 2014; 6(8): 2247-73.
[http://dx.doi.org/10.3390/polym6082247]

[75] Zhang C, Garrison TF, Madbouly SA, Kessler MR. Recent advances in vegetable oil-based polymers and their composites. Prog Polym Sci 2017; 71: 91-143.
[http://dx.doi.org/10.1016/j.progpolymsci.2016.12.009]

[76] Guzelgulgen M, Ozkendir-Inanc D, Yildiz UH, Arslan-Yildiz A. Glucuronoxylan-based quince seed hydrogel: A promising scaffold for tissue engineering applications. Int J Biol Macromol 2021; 180: 729-38.
[http://dx.doi.org/10.1016/j.ijbiomac.2021.03.096] [PMID: 33757854]

[77] Narayanan KB, Zo SM, Han SS. Novel biomimetic chitin-glucan polysaccharide nano/microfibrous fungal-scaffolds for tissue engineering applications. Int J Biol Macromol 2020; 149: 724-31.
[http://dx.doi.org/10.1016/j.ijbiomac.2020.01.276] [PMID: 32004611]

[78] de Oliveira AC, Sabino RM, Souza PR, *et al.* Chitosan/gellan gum ratio content into blends modulates the scaffolding capacity of hydrogels on bone mesenchymal stem cells. Mater Sci Eng C 2020; 106: 110258.
[http://dx.doi.org/10.1016/j.msec.2019.110258] [PMID: 31753363]

[79] Tong Y, Wang Z, Xiao Y, *et al. In situ* forming and reversibly cross-linkable hydrogels based on copolypept(o)ides and polysaccharides. ACS Appl Bio Mater 2019; 2(10): 4545-56.
[http://dx.doi.org/10.1021/acsabm.9b00668] [PMID: 35021414]

[80] Bazmandeh AZ, Mirzaei E, Ghasemi Y, Kouhbanani MAJ. Hyaluronic acid coated electrospun chitosan-based nanofibers prepared by simultaneous stabilizing and coating. Int J Biol Macromol 2019; 138: 403-11.
[http://dx.doi.org/10.1016/j.ijbiomac.2019.07.107] [PMID: 31326513]

[81] Gao Y, Jin X. Dual crosslinked methacrylated alginate hydrogel micron fibers and tissue constructs for cell biology. Mar Drugs 2019; 17(10): 557.
[http://dx.doi.org/10.3390/md17100557] [PMID: 31569386]

[82] Pourjavadi A, Dorcudian M, Ahadpour A, Azari S. Injectable chitosan/κ-carrageenan hydrogel designed with au nanoparticles: A conductive scaffold for tissue engineering demands. Int J Biol Macromol 2019; 126: 310-7.
[http://dx.doi.org/10.1016/j.ijbiomac.2018.11.256] [PMID: 30502431]

[83] You Y, Xie Y, Jiang Z. Injectable and biocompatible chitosan-alginic acid hydrogels. Biomed Mater 2019; 14(2): 025010.
[http://dx.doi.org/10.1088/1748-605X/aaff3d] [PMID: 30650388]

[84] Maciel JS, Azevedo S, Correia CR, *et al.* Oxidized cashew gum scaffolds for tissue engineering. Macromol Mater Eng 2019; 304(3): 1800574.
[http://dx.doi.org/10.1002/mame.201800574]

[85] Bombaldi de Souza RF, Bombaldi de Souza FC, Rodrigues C, *et al.* Mechanically-enhanced

polysaccharide-based scaffolds for tissue engineering of soft tissues. Mater Sci Eng C 2019; 94: 364-75.
[http://dx.doi.org/10.1016/j.msec.2018.09.045] [PMID: 30423719]

[86]   Gomes S, Rodrigues G, Martins G, Henriques C, Silva JC. Evaluation of nanofibrous scaffolds obtained from blends of chitosan, gelatin and polycaprolactone for skin tissue engineering. Int J Biol Macromol 2017; 102: 1174-85.
[http://dx.doi.org/10.1016/j.ijbiomac.2017.05.004] [PMID: 28487195]

[87]   Begum ERA, Rajaiah S, Bhavani K, Devi M, Karthika K, Gowri Priya C. Evaluation of extracted chitosan from portunus pelagicus for the preparation of chitosan alginate blend scaffolds. J Polym Environ 2017; 25(3): 578-85.
[http://dx.doi.org/10.1007/s10924-016-0834-z]

[88]   Wang Y, Qian J, Zhao N, Liu T, Xu W, Suo A. Novel hydroxyethyl chitosan/cellulose scaffolds with bubble-like porous structure for bone tissue engineering. Carbohydr Polym 2017; 167: 44-51.
[http://dx.doi.org/10.1016/j.carbpol.2017.03.030] [PMID: 28433176]

[89]   Kirdponpattara S, Khamkeaw A, Sanchavanakit N, Pavasant P, Phisalaphong M. Structural modification and characterization of bacterial cellulose–alginate composite scaffolds for tissue engineering. Carbohydr Polym 2015; 132: 146-55.
[http://dx.doi.org/10.1016/j.carbpol.2015.06.059] [PMID: 26256335]

[90]   Bueno VB, Takahashi SH, Catalani LH, de Torresi SIC, Petri DFS. Biocompatible xanthan/polypyrrole scaffolds for tissue engineering. Mater Sci Eng C 2015; 52: 121-8.
[http://dx.doi.org/10.1016/j.msec.2015.03.023] [PMID: 25953548]

[91]   Campbell FC. Thermoset resins: the glue that holds the strings together. In: Campbell FC, Ed. Manufacturing processes for advanced composites. Elsevier Science 2004; pp. 63-101.
[http://dx.doi.org/10.1016/B978-185617415-2/50004-6]

[92]   Fang S, Li F, Liu J, *et al.* Rubber-reinforced rubbers toward the combination of high reinforcement and low energy loss. Nano Energy 2021; 83: 105822.
[http://dx.doi.org/10.1016/j.nanoen.2021.105822]

[93]   Li Y, Zhu J, Cheng H, *et al.* Developments of advanced electrospinning techniques: a critical review. Adv Mater Technol 2021; 6(11): 2100410.
[http://dx.doi.org/10.1002/admt.202100410]

[94]   Taylor GI, Van Dyke MD. Electrically driven jets. Proceed Royal Societ A. Mathematical and Physical Sciences 1969; 313(1515): 453-75.

[95]   Rosman N, Wan Salleh WN, Jamalludin MR, *et al.* Electrospinning parameters evaluation of PVDF-ZnO/Ag$_2$CO$_3$/Ag$_2$O composite nanofiber affect on porosity by using response surface methodology. Mater Today Proc 2021; 46: 1824-30.
[http://dx.doi.org/10.1016/j.matpr.2020.11.847]

[96]   Aslam Khan MU, Abd Razak SI, Al Arjan WS, *et al.* Recent advances in biopolymeric composite materials for tissue engineering and regenerative medicines: a review. Molecules 2021; 26(3): 619.
[http://dx.doi.org/10.3390/molecules26030619] [PMID: 33504080]

[97]   Nyamweya NN. Applications of polymer blends in drug delivery. Future Journal of Pharmaceutical Sciences 2021; 7(1): 18.
[http://dx.doi.org/10.1186/s43094-020-00167-2]

[98]   Elakkiya T, Malarvizhi G, Rajiv S, Natarajan TS. Curcumin loaded electrospun *Bombyx mori* silk nanofibers for drug delivery. Polym Int 2014; 63(1): 100-5.
[http://dx.doi.org/10.1002/pi.4499]

[99]   Jeong L, Kim MH, Jung J-Y, Min BM, Park WH. Effect of silk fibroin nanofibers containing silver sulfadiazine on wound healing. Int J Nanomedicine 2014; 9: 5277-87.
[PMID: 25484581]

[100] Mohsenzadeh M, Alizadeh-Sani M, Maleki M, Azizi-lalabadi M, Rezaeian-Doloei R. Fabrication of biocomposite films based on sodium caseinate reinforced with gellan and Persian gums and evaluation of physicomechanical and morphology properties. Food Sci Technol (Campinas) 2021; 18(113): 187-96.
[http://dx.doi.org/10.52547/fsct.18.113.187]

[101] Hao Y, Zheng W, Sun Z, *et al.* Marine polysaccharide-based composite hydrogels containing fucoidan: Preparation, physicochemical characterization, and biocompatible evaluation. Int J Biol Macromol 2021; 183: 1978-86.
[http://dx.doi.org/10.1016/j.ijbiomac.2021.05.190] [PMID: 34087304]

[102] Zhu F. Polysaccharide based films and coatings for food packaging: Effect of added polyphenols. Food Chem 2021; 359: 129871.
[http://dx.doi.org/10.1016/j.foodchem.2021.129871] [PMID: 34023728]

[103] Islam HBMZ, Susan MABH, Imran AB. High-strength potato starch/hectorite clay-based nanocomposite film: synthesis and characterization. Iran Polym J 2021; 30(5): 513-21.
[http://dx.doi.org/10.1007/s13726-021-00907-y]

[104] Bof MJ, Locaso DE, García MA. Corn starch-chitosan proportion affects biodegradable film performance for food packaging purposes. Stärke 2021; 73(5-6): 2000104.
[http://dx.doi.org/10.1002/star.202000104]

[105] Sommer A, Staroszczyk H, Sinkiewicz I, Bruździak P. Preparation and characterization of films based on disintegrated bacterial cellulose and montmorillonite. J Polym Environ 2021; 29(5): 1526-41.
[http://dx.doi.org/10.1007/s10924-020-01968-5]

[106] Zannini D, Dal Poggetto G, Malinconico M, Santagata G, Immirzi B. Citrus pomace biomass as a source of pectin and lignocellulose fibers: from waste to upgraded biocomposites for mulching applications. Polymers (Basel) 2021; 13(8): 1280.
[http://dx.doi.org/10.3390/polym13081280] [PMID: 33919976]

[107] Nwabor OF, Singh S, Paosen S, Vongkamjan K, Voravuthikunchai SP. Enhancement of food shelf life with polyvinyl alcohol-chitosan nanocomposite films from bioactive Eucalyptus leaf extracts. Food Biosci 2020; 36: 100609.
[http://dx.doi.org/10.1016/j.fbio.2020.100609]

[108] Singh S, Nwabor OF, Syukri DM, Voravuthikunchai SP. Chitosan-poly(vinyl alcohol) intelligent films fortified with anthocyanins isolated from *Clitoria ternatea* and *Carissa carandas* for monitoring beverage freshness. Int J Biol Macromol 2021; 182: 1015-25.
[http://dx.doi.org/10.1016/j.ijbiomac.2021.04.027] [PMID: 33839180]

[109] Guo Y, Zhang B, Zhao S, Qiao D, Xie F. Plasticized starch/agar composite films: processing, morphology, structure, mechanical properties and surface hydrophilicity. Coatings 2021; 11(3): 311.
[http://dx.doi.org/10.3390/coatings11030311]

[110] Xu J, Sagnelli D, Faisal M, *et al.* Amylose/cellulose nanofiber composites for all-natural, fully biodegradable and flexible bioplastics. Carbohydr Polym 2021; 253: 117277.
[http://dx.doi.org/10.1016/j.carbpol.2020.117277] [PMID: 33278948]

[111] He Y, Zhao W, Dong Z, *et al.* A biodegradable antibacterial alginate/carboxymethyl chitosan/Kangfuxin sponges for promoting blood coagulation and full-thickness wound healing. Int J Biol Macromol 2021; 167: 182-92.
[http://dx.doi.org/10.1016/j.ijbiomac.2020.11.168] [PMID: 33259842]

[112] Cakıroglu K, Dervisoglu M, Gul O. Development and characterization of black mulberry (*Morus nigra*) pekmez (molasses) composite films based on alginate and pectin. J Texture Stud 2020; 51(5): 800-9.
[http://dx.doi.org/10.1111/jtxs.12528] [PMID: 32358987]

[113] Wang H, Kong L, Ziegler GR. Fabrication of starch - Nanocellulose composite fibers by

electrospinning. Food Hydrocoll 2019; 90: 90-8.
[http://dx.doi.org/10.1016/j.foodhyd.2018.11.047]

[114]   Zheng J, Yang Q, Shi X, Xie Z, Hu J, Liu Y. Effects of preparation parameters on the properties of the crosslinked pectin nanofiber mats. Carbohydr Polym 2021; 269: 118314.
[http://dx.doi.org/10.1016/j.carbpol.2021.118314] [PMID: 34294328]

[115]   Shokraei S, Mirzaei E, Shokraei N, Derakhshan MA, Ghanbari H, Faridi-Majidi R. Fabrication and characterization of chitosan/kefiran electrospun nanofibers for tissue engineering applications. J Appl Polym Sci 2021; 138(24): 50547.
[http://dx.doi.org/10.1002/app.50547]

[116]   Poddar S, Agarwal PS, Sahi AK, Varshney N, Vajanthri KY, Mahto SK. Fabrication and characterization of electrospun psyllium husk□based nanofibers for tissue regeneration. J Appl Polym Sci 2021; 138(24): 50569.
[http://dx.doi.org/10.1002/app.50569]

[117]   Dodero A, Donati I, Scarfi S, *et al.* Effect of sodium alginate molecular structure on electrospun membrane cell adhesion. Mater Sci Eng C 2021; 124: 112067.
[http://dx.doi.org/10.1016/j.msec.2021.112067] [PMID: 33947560]

[118]   Toullec C, Le Bideau J, Geoffroy V, *et al.* curdlan–chitosan electrospun fibers as potential scaffolds for bone regeneration. Polymers (Basel) 2021; 13(4): 526.
[http://dx.doi.org/10.3390/polym13040526] [PMID: 33578913]

[119]   Mishra P, Gupta P, Pruthi V. Cinnamaldehyde incorporated gellan/PVA electrospun nanofibers for eradicating *Candida* biofilm. Mater Sci Eng C 2021; 119: 111450.
[http://dx.doi.org/10.1016/j.msec.2020.111450] [PMID: 33321588]

[120]   Jia X, Qin Z, Xu J, Kong B, Liu Q, Wang H. Preparation and characterization of pea protein isolate-pullulan blend electrospun nanofiber films. Int J Biol Macromol 2020; 157: 641-7.
[http://dx.doi.org/10.1016/j.ijbiomac.2019.11.216] [PMID: 31786299]

[121]   Yang Y, Xie B, Liu Q, Kong B, Wang H. Fabrication and characterization of a novel polysaccharide based composite nanofiber films with tunable physical properties. Carbohydr Polym 2020; 236: 116054.
[http://dx.doi.org/10.1016/j.carbpol.2020.116054] [PMID: 32172869]

[122]   Mamun A, Trabelsi M, Klöcker M, Lukas Storck J, Böttjer R, Sabantina L. Needleless electrospun polyacrylonitrile/konjac glucomannan nanofiber mats. J Eng Fibers Fabrics 2020; 15
[http://dx.doi.org/10.1177/1558925020964806]

[123]   Serife Safak OV. Nilufer Cinkilic, Esra Karaca. *In vitro* evaluation of electrospun polysaccharide based nanofibrous mats as surgical adhesion barriers. Tekstil Ve Konfeksiyon 2020; 30(2): 99-107.

[124]   Perez-Puyana V, Felix M, Cabrera L, Romero A, Guerrero A. Development of gelatin/chitosan membranes with controlled microstructure by electrospinning. Iran Polym J 2019; 28(11): 921-31.
[http://dx.doi.org/10.1007/s13726-019-00755-x]

[125]   Martín-Alfonso JE, Cuadri AA, Franco JM. Development and characterization of novel fibers based on potato protein/polyethylene oxide through electrospinning. Fibers Polym 2019; 20(8): 1586-93.
[http://dx.doi.org/10.1007/s12221-019-1258-x]

[126]   Petrova VA, Chernyakov DD, Poshina DN, *et al.* Electrospun bilayer chitosan/hyaluronan material and its compatibility with mesenchymal stem cells. Materials (Basel) 2019; 12(12): 2016.
[http://dx.doi.org/10.3390/ma12122016] [PMID: 31238491]

[127]   Faralli A, Shekarforoush E, Ajalloueian F, Mendes AC, Chronakis IS. *In vitro* permeability enhancement of curcumin across Caco-2 cells monolayers using electrospun xanthan-chitosan nanofibers. Carbohydr Polym 2019; 206: 38-47.
[http://dx.doi.org/10.1016/j.carbpol.2018.10.073] [PMID: 30553335]

[128]   Eze FN, Jayeoye TJ, Singh S. Fabrication of intelligent pH-sensing films with antioxidant potential for

monitoring shrimp freshness *via* the fortification of chitosan matrix with broken Riceberry phenolic extract. Food Chem 2022; 366: 130574.
[http://dx.doi.org/10.1016/j.foodchem.2021.130574] [PMID: 34303209]

[129] Liu T, Liu L, Gong X, Chi F, Ma Z. Fabrication and comparison of active films from chitosan incorporating different spice extracts for shelf life extension of refrigerated pork. Lebensm Wiss Technol 2021; 135: 110181.
[http://dx.doi.org/10.1016/j.lwt.2020.110181]

[130] Yong H, Liu Y, Yun D, Zong S, Jin C, Liu J. Chitosan films functionalized with different hydroxycinnamic acids: preparation, characterization and application for pork preservation. Foods 2021; 10(3): 536.
[http://dx.doi.org/10.3390/foods10030536] [PMID: 33807529]

[131] Wang Y, Zeng L, Song W, Liu J. Influencing factors and drug application of iontophoresis in transdermal drug delivery: an overview of recent progress. Drug Deliv Translat Res 2021; pp. 1-12.

[132] Prausnitz MR. Microneedles for transdermal drug delivery. Adv Drug Deliv Rev 2004; 56(5): 581-7.
[http://dx.doi.org/10.1016/j.addr.2003.10.023] [PMID: 15019747]

[133] v Badhe R, Adkine D, Godse A. Development of polylactic acid and bovine serum albumin layered coated chitosan microneedles using novel bees wax mould. Turkish Journal of Pharmaceutical Sciences 2021; 18(3): 367-75.
[http://dx.doi.org/10.4274/tjps.galenos.2020.47897] [PMID: 34157828]

[134] Ling MH, Chen MC. Dissolving polymer microneedle patches for rapid and efficient transdermal delivery of insulin to diabetic rats. Acta Biomater 2013; 9(11): 8952-61.
[http://dx.doi.org/10.1016/j.actbio.2013.06.029] [PMID: 23816646]

[135] Zhang XP, Wang BB, Li WX, Fei WM, Cui Y, Guo XD. *In vivo* safety assessment, biodistribution and toxicology of polyvinyl alcohol microneedles with 160-day uninterruptedly applications in mice. Eur J Pharm Biopharm 2021; 160: 1-8.
[http://dx.doi.org/10.1016/j.ejpb.2021.01.005] [PMID: 33484865]

[136] Zare MR, Khorram M, Barzegar S, *et al.* Dissolvable carboxymethyl cellulose/polyvinylpyrrolidone microneedle arrays for transdermal delivery of Amphotericin B to treat cutaneous leishmaniasis. Int J Biol Macromol 2021; 182: 1310-21.
[http://dx.doi.org/10.1016/j.ijbiomac.2021.05.075] [PMID: 34000308]

[137] Yu W, Jiang G, Zhang Y, Liu D, Xu B, Zhou J. Polymer microneedles fabricated from alginate and hyaluronate for transdermal delivery of insulin. Mater Sci Eng C 2017; 80: 187-96.
[http://dx.doi.org/10.1016/j.msec.2017.05.143] [PMID: 28866156]

[138] Fonseca DFS, Costa PC, Almeida IF, *et al.* Pullulan microneedle patches for the efficient transdermal administration of insulin envisioning diabetes treatment. Carbohydr Polym 2020; 241: 116314.
[http://dx.doi.org/10.1016/j.carbpol.2020.116314] [PMID: 32507191]

[139] Olatunji O, Igwe CC, Ahmed AS, Alhassan DOA, Asieba GGO, Diganta BD. Microneedles from fish scale biopolymer. J Appl Polym Sci 2014; 131(12): n/a.
[http://dx.doi.org/10.1002/app.40377]

[140] Medhi P, Olatunji O, Nayak A, *et al.* Lidocaine-loaded fish scale-nanocellulose biopolymer composite microneedles. AAPS PharmSciTech 2017; 18(5): 1488-94.
[http://dx.doi.org/10.1208/s12249-017-0758-5] [PMID: 28353171]

[141] Olatunji O, Olsson R. Microneedles from fishscale-nanocellulose blends using low temperature mechanical press method. Pharmaceutics 2015; 7(4): 363-78.
[http://dx.doi.org/10.3390/pharmaceutics7040363] [PMID: 26404358]

[142] Michael AL, Danielle RB, John CR, Ronald AS, Jeremiah J. Biodegradable 3D printed polymer microneedles for transdermal drug delivery. Lab on a Chip 2018; 18: 1223-230. Gassensmith Olatunji O, Igwe CC, Ahmed AS, Alhassan DOA, Asieba GO, Diganta BD. Microneedles from fish scale

biopolymer. J Appl Polym Sci 2014; 131(12)

[143]   Olatunji O, Olubowale M, Okereke C. Microneedle-assisted transdermal delivery of acetylsalicylic acid (aspirin) from biopolymer films extracted from fish scales. Polym Bull 2018; 75(9): 4103-15.
[http://dx.doi.org/10.1007/s00289-017-2254-1]

[144]   Olatunji O, Denloye A. Production of hydrogel microneedles from fish scale biopolymer. J Polym Environ 2019; 27(6): 1252-8.
[http://dx.doi.org/10.1007/s10924-019-01426-x]

[145]   Liu T, Jiang G, Song G, Sun Y, Zhang X, Zeng Z. Fabrication of rapidly separable microneedles for transdermal delivery of metformin on diabetic rats. J Pharm Sci 2021; 110(8): 3004-10.
[http://dx.doi.org/10.1016/j.xphs.2021.04.009] [PMID: 33878323]

[146]   Vora LK, Courtenay AJ, Tekko IA, Larrañeta E, Donnelly RF. Pullulan-based dissolving microneedle arrays for enhanced transdermal delivery of small and large biomolecules. Int J Biol Macromol 2020; 146: 290-8.
[http://dx.doi.org/10.1016/j.ijbiomac.2019.12.184] [PMID: 31883883]

[147]   Shah V, Choudhury BK. Fabrication, physicochemical characterization, and performance evaluation of biodegradable polymeric microneedle patch system for enhanced transcutaneous flux of high molecular weight therapeutics. AAPS PharmSciTech 2017; 18(8): 2936-48.
[http://dx.doi.org/10.1208/s12249-017-0774-5] [PMID: 28432615]

[148]   Zhu M, Liu Y, Jiang F, Cao J, Kundu SC, Lu S. Combined silk fibroin microneedles for insulin delivery. ACS Biomater Sci Eng 2020; 6(6): 3422-9.
[http://dx.doi.org/10.1021/acsbiomaterials.0c00273] [PMID: 33463180]

[149]   Wang C, Liu S, Xu J, *et al.* Dissolvable microneedles based on Panax notoginseng polysaccharide for transdermal drug delivery and skin dendritic cell activation. Carbohydr Polym 2021; 268: 118211.
[http://dx.doi.org/10.1016/j.carbpol.2021.118211] [PMID: 34127215]

[150]   Hu L, Liao Z, Hu Q, Maffucci KG, Qu Y. Novel Bletilla striata polysaccharide microneedles: Fabrication, characterization, and in vitro transcutaneous drug delivery. Int J Biol Macromol 2018; 117: 928-36.
[http://dx.doi.org/10.1016/j.ijbiomac.2018.05.097] [PMID: 29775714]

[151]   Yalcintas EP, Ackerman DS, Korkmaz E, *et al.* Analysis of *in vitro* cytotoxicity of carbohydrate-based materials used for dissolvable microneedle arrays. Pharm Res 2020; 37(3): 33.
[http://dx.doi.org/10.1007/s11095-019-2748-7] [PMID: 31942659]

[152]   Zhang Y, Jiang G, Weijiang Y, Liu D, Xu B. Microneedles fabricated from alginate and maltose for transdermal delivery of insulin on diabetic rats. Mater Sci Eng C 2017; 85.
[PMID: 29407146]

[153]   Roszek B, De Jong W, Geertsma RE. Nanotechnology in medical applications: state-of-the-art in materials and devices. RIVM Report 2005.

[154]   Dekker F, Kool L, Bunschoten A, Velders AH, Saggiomo V. Syntheses of gold and silver dichroic nanoparticles; looking at the Lycurgus cup colors. Chem Teach Int 2021; 3(1)

[155]   Harish Kumar KNV. Himangshu Bhowmik, Anuttam Kuila. Metallic nanoparticle: a review. Biomed J Sci Tech Res 2018; 4(2): 3765-75.

[156]   Alavi M, Rai M. Recent progress in nanoformulations of silver nanoparticles with cellulose, chitosan, and alginic acid biopolymers for antibacterial applications. Appl Microbiol Biotechnol 2019; 103(21-22): 8669-76.
[http://dx.doi.org/10.1007/s00253-019-10126-4] [PMID: 31522283]

[157]   Teixeira MA, Paiva MC, Amorim MTP, Felgueiras HP. Felgueiras, Helena P. Electrospun nanocomposites containing cellulose and its derivatives modified with specialized biomolecules for an enhanced wound healing. Nanomaterials (Basel) 2020; 10(3): 557.
[http://dx.doi.org/10.3390/nano10030557]

[158]   Bhanja SK, Samanta SK, Mondal B, *et al.* Green synthesis of Ag@Au bimetallic composite nanoparticles using a polysaccharide extracted from *Ramaria botrytis* mushroom and performance in catalytic reduction of 4-nitrophenol and antioxidant, antibacterial activity. Environ Nanotechnol Monit Manag 2020; 14: 100341.
[http://dx.doi.org/10.1016/j.enmm.2020.100341]

[159]   El-Rafie HM, El-Rafie MH, Zahran MK. Green synthesis of silver nanoparticles using polysaccharides extracted from marine macro algae. Carbohydr Polym 2013; 96(2): 403-10.
[http://dx.doi.org/10.1016/j.carbpol.2013.03.071] [PMID: 23768580]

[160]   Yugay YA, Usoltseva RV, Silant'ev VE, *et al.* Synthesis of bioactive silver nanoparticles using alginate, fucoidan and laminaran from brown algae as a reducing and stabilizing agent. Carbohydr Polym 2020; 245: 116547.
[http://dx.doi.org/10.1016/j.carbpol.2020.116547] [PMID: 32718640]

[161]   Hileuskaya K, Ladutska A, Kulikouskaya V, *et al.* 'Green' approach for obtaining stable pectin-capped silver nanoparticles: Physico-chemical characterization and antibacterial activity. Colloids Surf A Physicochem Eng Asp 2020; 585: 124141.
[http://dx.doi.org/10.1016/j.colsurfa.2019.124141]

[162]   Brito TK, Silva Viana RL, Gonçalves Moreno CJ, *et al.* Synthesis of Silver Nanoparticle Employing Corn Cob Xylan as a Reducing Agent with Anti-*Trypanosoma cruzi* Activity. Int J Nanomedicine 2020; 15: 965-79.
[http://dx.doi.org/10.2147/IJN.S216386] [PMID: 32103950]

[163]   Sivasankar P, Seedevi P, Poongodi S, *et al.* Characterization, antimicrobial and antioxidant property of exopolysaccharide mediated silver nanoparticles synthesized by Streptomyces violaceus MM72. Carbohydr Polym 2018; 181: 752-9.
[http://dx.doi.org/10.1016/j.carbpol.2017.11.082] [PMID: 29254032]

[164]   Selvakumar R, Aravindh S, Ashok AM, Balachandran YL. A facile synthesis of silver nanoparticle with SERS and antimicrobial activity using *Bacillus subtilis* exopolysaccharides. J Exp Nanosci 2014; 9(10): 1075-87.
[http://dx.doi.org/10.1080/17458080.2013.778425]

[165]   Sanyasi S, Majhi RK, Kumar S, *et al.* Polysaccharide-capped silver Nanoparticles inhibit biofilm formation and eliminate multi-drug-resistant bacteria by disrupting bacterial cytoskeleton with reduced cytotoxicity towards mammalian cells. Sci Rep 2016; 6(1): 24929.
[http://dx.doi.org/10.1038/srep24929] [PMID: 27125749]

[166]   Jayeoye TJ, Eze FN, Singh S, Olatunde OO, Benjakul S, Rujiralai T. Synthesis of gold nanoparticles/polyaniline boronic acid/sodium alginate aqueous nanocomposite based on chemical oxidative polymerization for biological applications. Int J Biol Macromol 2021; 179: 196-205.
[http://dx.doi.org/10.1016/j.ijbiomac.2021.02.199] [PMID: 33675826]

[167]   Jayeoye TJ, Sirimahachai U, Rujiralai T. Sensitive colorimetric detection of ascorbic acid based on seed mediated growth of sodium alginate reduced/stabilized gold nanoparticles. Carbohydr Polym 2021; 255: 117376.
[http://dx.doi.org/10.1016/j.carbpol.2020.117376] [PMID: 33436207]

[168]   Jayeoye TJ, Rujiralai T. Green, *in situ* fabrication of silver/poly(3-aminophenyl boronic acid)/sodium alginate nanogel and hydrogen peroxide sensing capacity. Carbohydr Polym 2020; 246: 116657.
[http://dx.doi.org/10.1016/j.carbpol.2020.116657] [PMID: 32747289]

[169]   Khalifa Z, Zahran M, A-H Zahran M, Azzem MA. Mucilage-capped silver nanoparticles for glucose electrochemical sensing and fuel cell applications. RSC Advances 2020; 10(62): 37675-82.
[http://dx.doi.org/10.1039/D0RA07359H] [PMID: 35515185]

[170]   Garnica-Romo MG, Coria-Caballero V, Tranquilino-Rodríguez E, *et al.* Ecological method for the synthesis, characterization and antimicrobial effect of silver nanoparticles produced and stabilized with a mixture of mucilage/proteins extracted from flaxseed. J Inorg Organomet Polym Mater 2021; 31(8):

3406-15.
[http://dx.doi.org/10.1007/s10904-021-01968-5]

[171] Prasad AR, Garvasis J, Oruvil SK, Joseph A. Bio-inspired green synthesis of zinc oxide nanoparticles using Abelmoschus esculentus mucilage and selective degradation of cationic dye pollutants. J Phys Chem Solids 2019; 127: 265-74.
[http://dx.doi.org/10.1016/j.jpcs.2019.01.003]

[172] Doppalapudi S, Jain A, Khan W, Domb AJ. Biodegradable polymers-an overview. Polym Adv Technol 2014; 25(5): 427-35.
[http://dx.doi.org/10.1002/pat.3305]

CHAPTER 7

Potential Application of Biopolymers as Biodegradable Plastic

Abstract: Today, on average, we produce yearly about 300 million tons of plastic waste, equivalent to the entire human population weight around the globe. The single-use plastics and plastic products are produced using high molecular weight polymers in combination with additives that could not be completely reprocessed. So far, only 9% of overall plastic waste produced has been recycled and around 12% has been incinerated while the remaining 79% has been accumulated as debris in terrestrial and aquatic environments. Biodegradability and recycling of plastics depend on various physicochemical properties including molecular weight, hydrogen bonding, Van der Waals forces, and electrostatic forces. Moreover, biodegradability depends on macromolecular chain distresses that not only affect polymer aggregates but also affect the structural and functional properties of plastic products. However, due to unlimited production and utility with distressing effects on the environment, it is deemed necessary to replace such non-biodegradable polymers used in the fabrication of plastics with biodegradable polymers. The use of biodegradable polymers in the fabrication of plastic products is a creative way to resolve the plastic disposal problem. In this chapter, a brief overview has been presented on the fabrication of biodegradable plastic using biopolymers to reduce its detrimental effects on the environment.

Keywords: Bio-based polymer, Bioplastic, Plastic, Plastic-packaging, Thermoplastic.

INTRODUCTION

Today nearly everyone globally every day comes in contact with plastic or products packed with plastics, which are inexpensive, gauzy, and durable materials. The term plastic was derived from *Plasticus* a Latin term and *Plastikos* a Greek word that was common in practice during the 17th century for molded materials [1]. Plastics are a pivotal material in modern life and are popularly used in water supply, food packing, and in various medical devices including surgical equipment including drips, aseptic medical packaging, and blister packs for pills [2]. Although plastic has various benefits, however, its bioburden became obvious day by day due to its short first-use cycle. Currently, plastic packaging represents 26% of the total produced volume of plastic with a direct economic boost that significantly contributes to productivity resources [3]. The first use of synthetic

Sudarshan Singh & Warangkana Chunglok

plastic was recorded in 1907 as Bakelite in Belgium [4]. While the commercial production of plastic materials started during the 1950s with remarkable global annual percentage production of 393 million metric tons including textile fabrics [5]. The mismanaged accumulation of plastic waste in the environment has raised a global concern. Statistical data indicates about 60 to 99 million metric tons of mismanaged plastic waste were produced globally in 2016 (Fig. **1**), and the Figs could triple to 155-265 metric tons per year by 2060 [6]. A previous report indicated that 32% of plastic packaging escapes from collection systems and 91% of the mismanaged plastic causes significant economic loss with detrimental effects on the ecosystem by clogging drainage and disposal to the ocean via rivers [7] (Fig. **2**). Moreover, approximately 4% of the fossil fuel extracted is consumed as raw material in the production of plastic [8]. The rise in the demand for plastic in packaging and construction materials indicates higher consumption of alternative energy during production, with an additional 4% increase in the emission of greenhouse gases [3]. In addition, the expected demand by 2050 for fossil fuel and other natural resources of energy as well as associated carbon emission by plastic producing industries may account for 20% of petroleum's consumable with 15% of carbon emission [7].

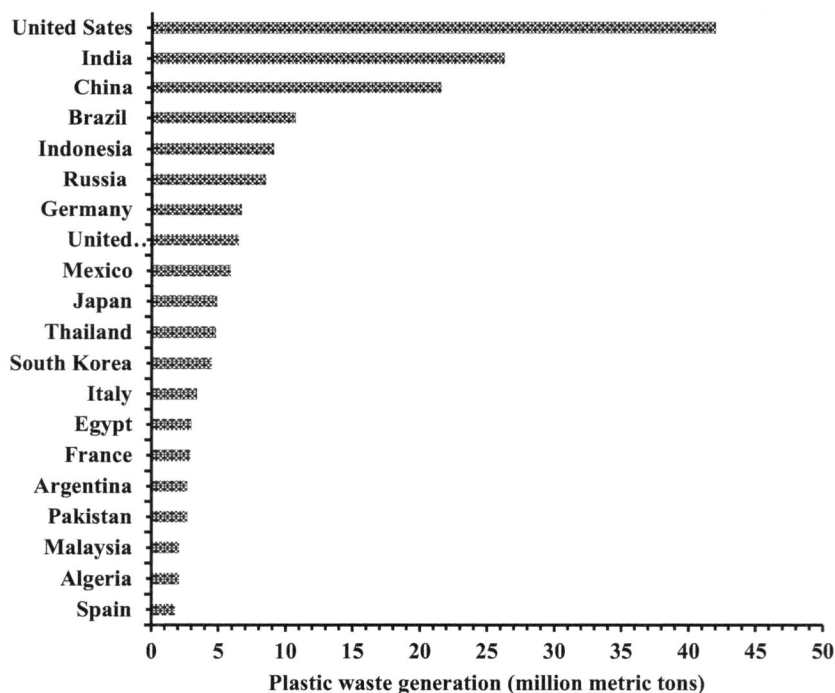

Fig. (1). The highest number of plastic waste was generated by the selected country in 2016 (in million metric tons).

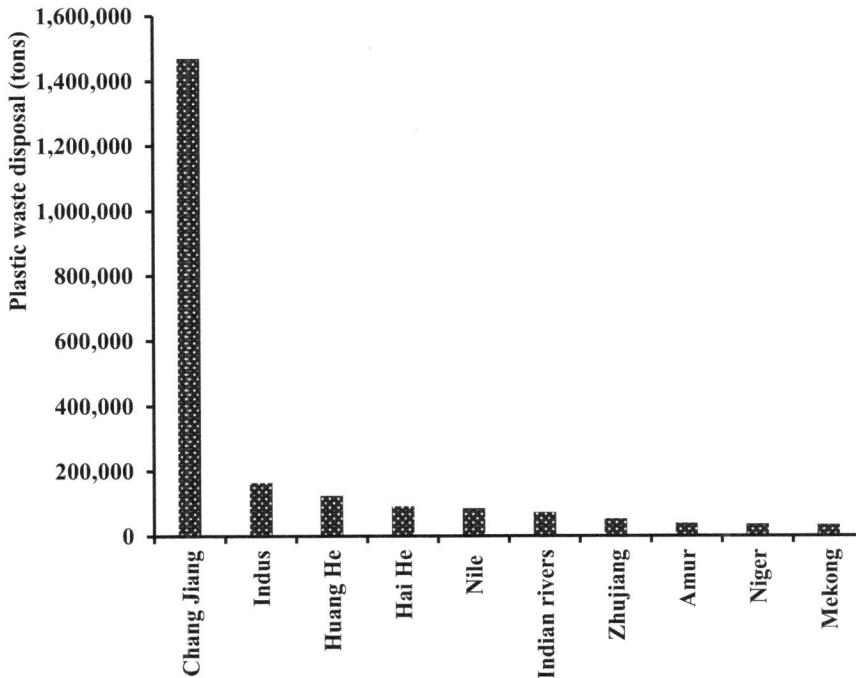

Fig. (2). Plastic waste disposal (tons) in top 10 rivers across the globe [9].

Presently 95% of plastic packaging materials costing 80-120 billion dollars annually are lost due to a short life cycle [3]. More than 4 decades after the launch of the universal plastic recycling process, only 14% of plastic packaging is collected with a 5% material loss during sorting and reprocessing causes a huge economic loss [10]. Moreover, the recycled plastics that do get recycled in general are used for a lower-value application that further cannot be reused. The recycling of plastic depends on the types of polymer used, package design, and product type. The plastic that is constituted of a single polymer is easier and more cost-effective to recycle, compared with complex plastic products. Different terminologies used for the plastic and plastic products during the recycling and recovery process are indicated in Table **1**.

Table 1. Different terminologies are used for plastic and plastic products during the recycling and recovery process.

ASTM D5033 Approach	ISO 15270 Correspondent	Analogous Expressions
Primary recycling	Mechanical recycling	Recycling with closed-loop
Secondary recycling	Mechanical recycling	Downgrading

(Table 1) cont.....

ASTM D5033 Approach	ISO 15270 Correspondent	Analogous Expressions
Tertiary recycling	Chemical recycling	Feed-stock recycling
Quaternary recycling	Recovery of energy	Valorization

Moreover, the plastic waste recycling process is presented in Fig. (**3**). Owing to the wide usage of plastic with additives, plastics pose several potential environmental and human health risks. Recently the concept of 5 "R" such as reduce, reuse, recycle, rethink, and restrain, has been implemented in the controlled production of plastics to address environmental and human health issues impersonated by plastic [11].

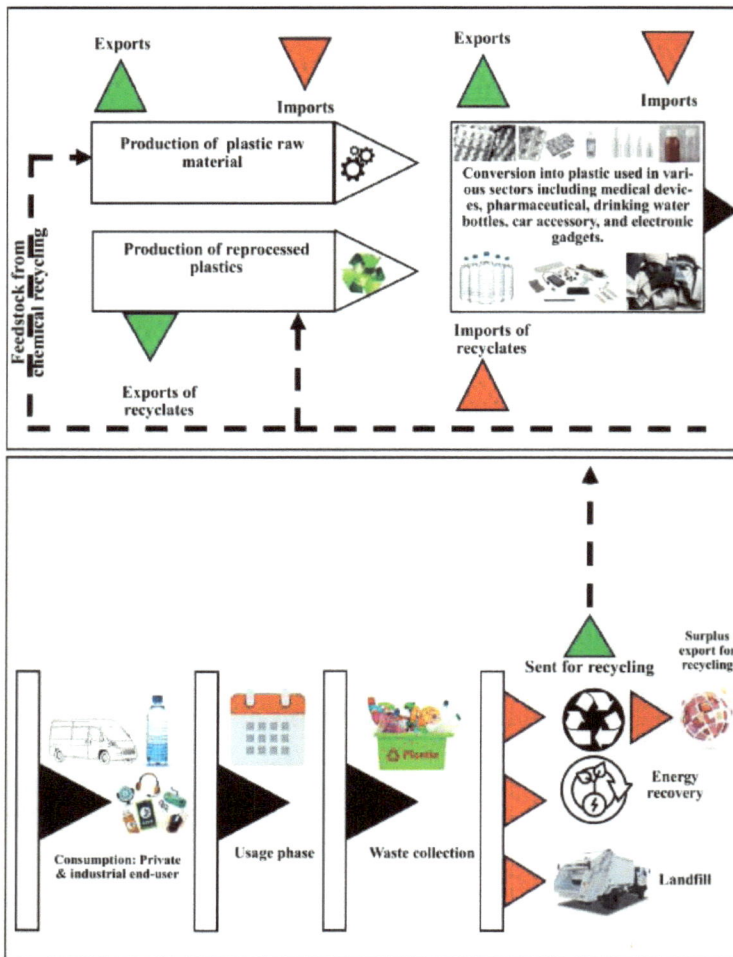

Fig. (3). Plastic waste recycling process [12].

Biopolymers as Biodegradable Plastics

The term bioplastic is often used to define the biodegradability of plastics, however not applicable to all types of plastics or plastic products. Generally, the polymers used in the production of plastics are extremely durable but non-biodegradable, and therefore, the majority of polymer plastics cause nasty effects to nature. Thus, there is an urgent need for the development of an alternative polymer or a material that can produce reusable plastic with biodegradability. The utilization of natural renewable polymers in the production of biodegradable plastic is a promising topic for research and industrial developments. The biodegradable polymers used in the fabrication of plastics are broadly classified as thermoplastics and thermosetting polymers. Thermoplastic polymers including polyethylene, polypropylene, *etc.* have the potential to be recycled mechanically (Fig. **4**), whereas, thermosetting polymers such as unsaturated polyester, and epoxy resin that cannot be mechanically recycled, might be due to permanent crosslinking of the materials during manufacturing (Fig. **5**) of plastic are largely used [13].

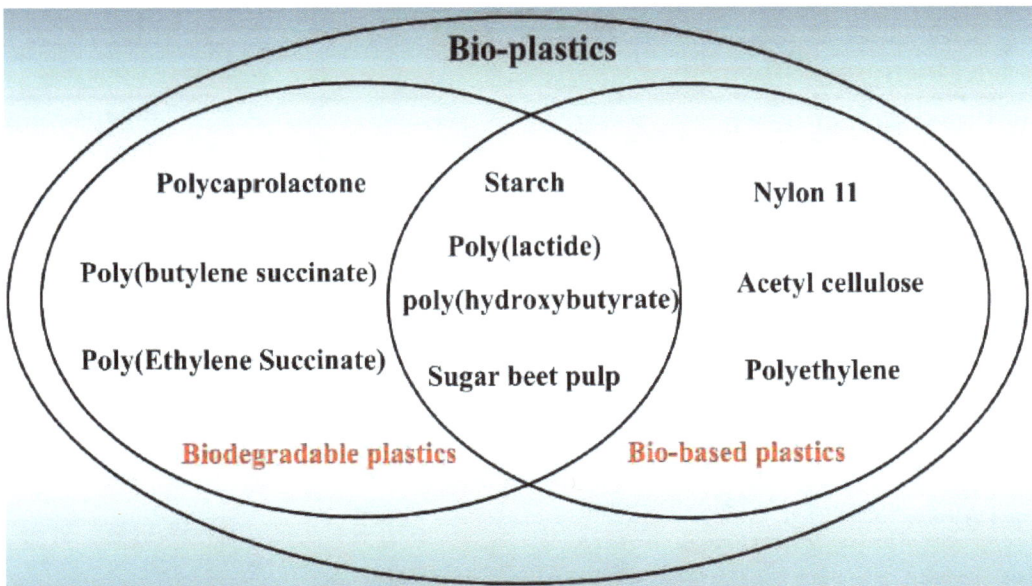

Fig. (4). Commonly used biopolymers in the development of bio-plastics.

In addition, several bio-based polymers have been produced through the polymerization of monomers obtained from natural resources such as polylactic acid which is obtained after the fermentation of sugarcane [14] or starch derived carbohydrates such as corn starch [15], bamboo bagasse [16], cassava bagasse

[17, 18], grape stalks [19], cogon grass fiber [20], and kraft fiber [21]. Starch-based polymers form an important class of bioplastic that has gained attention in the production of bioplastics. Starch is a completely biodegradable polysaccharide that consists of two macromolecules such as amylose and amylopectin [22]. The amylose is a linear polymer of D-glucose units attached with α-1,4-, and amylopectin, a highly branched polymer of D-glucose units attached with α-1,4- with branch points attached with α-1,6- [23]. Native starches are commercially processed with a plasticizer such as gelatin to produce thermoplastic starch with enhanced mechanical properties. Starch is renewable, readily biodegradable, easily modified both physically and chemically, and available in bulk in all parts of the world at low cost (estimated global production > 70 million tons in 2011), making it a very attractive raw material for the manufacturing of green plastics [24].

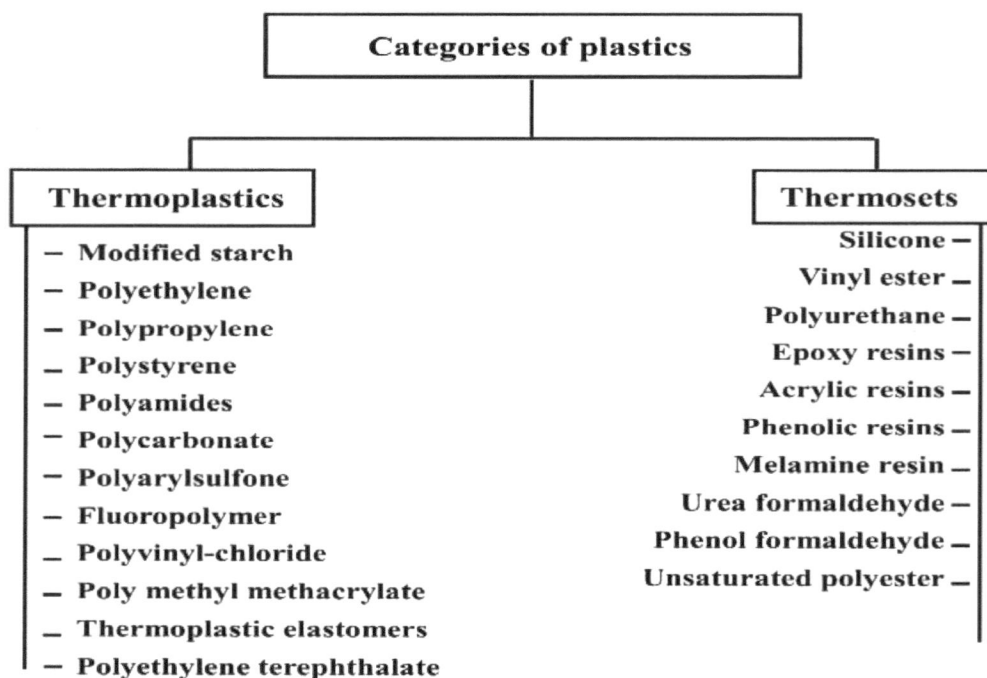

```
                      ┌──────────────────────────────────┐
                      │      Categories of plastics       │
                      └──────────────────────────────────┘

  ┌──────────────────────────┐          ┌──────────────────────────┐
  │     Thermoplastics       │          │        Thermosets         │
  └──────────────────────────┘          └──────────────────────────┘
    — Modified starch                       Silicone —
    — Polyethylene                          Vinyl ester —
    — Polypropylene                         Polyurethane —
    — Polystyrene                           Epoxy resins —
    — Polyamides                            Acrylic resins —
    — Polycarbonate                         Phenolic resins —
    — Polyarylsulfone                       Melamine resin —
    — Fluoropolymer                         Urea formaldehyde —
    — Polyvinyl-chloride                    Phenol formaldehyde —
    — Poly methyl methacrylate              Unsaturated polyester —
    — Thermoplastic elastomers
    — Polyethylene terephthalate
```

Fig. (5). Classification of polymers according to the manufacturing of various types of plastics.

Starch exists in a granular form, however, it loses its original structure to form a molten plastic state when subjected to shear forces at a temperature range of 90-180 °C in the presence of a plasticizer known as thermoplastic starch. The mechanical property of thermoplastic starch can be improved significantly by blending it with other polymers, fillers, and fibers. Both natural and synthetic polymers have been used, including cellulose, zein, natural rubber, polyvinyl alcohol, acrylate copolymers, polyethylene and ethylene copolymers, polyesters, and polyurethanes [25]. Despite several advantages with starch, it has a few drawbacks including moisture uptake with loss of mechanical properties. A recent report on the potential application of bio-based polymers in bioplastics manufacturing is presented in Table **2**.

Table 2. Potential application of bio-based polymers in bioplastics manufacturing.

Source of Bio-Based Polymer	Application in Bioplastics Manufacturing	Reference
Opuntia ficus indica	Improvement in plasticity and mechanical properties of thermoplastic film originating from potato starch and glycerol with the addition of mucilage.	[29]
Opuntia heliabravoana	Edible coating for the preservation of blackberries with thermoplastic starch indicated improved stability during storage.	[30]
Tamarindus indica, sugarcane bagasse, *Abelmoschus esculentus,* rice bran	Physiochemical characterization showed that natural gum isolated, except okra mucilage, could be used in the development of bio-thermoplastics.	[31, 32]
Natto foods are virtually produced using *Bacillus subtilis* (*natto*)	Poly-γ-glutamate a major compound derived from mucilage of natto, with hexadecyl pyridinium cation was transformed into thermoplastic materials with good thermal stability suggesting suitability for use in the manufacturing of soft plastic.	[33]
Potato starch	The combination of starch with potato peel phenolic extract enhanced the physio-mechanical properties of the thermoplastic film with improved packaging performance.	[34]
Sugarcane bagasse	Mixed-ester derivative of cellulose and hemicellulose with different chains showed significant mechanical properties that facilitate the fabrication of thermoplastics.	[35]
Soy protein	Soy protein and polyvinyl alcohol with montmorillonite bio-nano composite film demonstrated improvement in mechanical properties that could be used in the coating.	[36]
Arabinoxylan from underutilized wheat bran	Chemically modified arabinoxylan demonstrated reduced glass transition below the degradation temperature of native gum, suggesting its use as bio-based polysaccharides in film, packaging, and as substrates for stretchable electronic application.	[37]
Corn starch	Sucrose based ionic liquid crystals and chloride atoms as a supramolecular inducer to form thermoplastic corn starch granules showed potential application in green and sustainable packaging materials.	[38]

(Table 2) cont.....

Source of Bio-Based Polymer	Application in Bioplastics Manufacturing	Reference
Sugar beet pulp	Agro waste and plant engineered polysaccharides amylose with cellulose nanofibers blend composite could be used as an alternative to synthetic bioplastic.	[39]
Corn starch	Thermoplastic corn starch with bacterial cellulose sandwich composite prepared by extrusion compression molding showed promising alternatives for packaging of the hygienic product.	[40]
Maize starch	Thermoplastic starch and cellulose nanocrystals films with plasticizers developed using extrusion compression method indicated effective hydrogen bonding interaction and controlled barrier properties.	[41]
Potato starch	Starch blended with polylactide composite foam prepared using compression showed improved stiffness and mechanical properties with a reduction in water absorbent capacity.	[42]
Corn starch	Micro-particles of starch and chitosan cross-linked plasticized with glycerol demonstrated improvement in mechanical properties and toughness.	[43]
Dried henequen (*Agave fourcroydes*)	Thermoplastic starch reinforced with cellulose nanofiber composites showed improvement in mechanical properties and reduction in water vapor permeability.	[44]
Maize starch	Maize starch reinforced with cellulose nanocrystals and waxy starch, plasticized with glycerol bio-nano composite showed enhancement in mechanical properties whereas no significant difference in barrier properties.	[45]
Cassava starch	Thermoplastic starch film prepared using spray-dried potato starch with urea and glycerol as a plasticizer resulted in ductile and brittle film for urea and glycerol as plasticizer, respectively.	[46]
Saccharum officinarum fibers	Novel solid-state ball-milled cellulose in thermoplastic starch showed excellent improvement in structural and mechanical properties.	[47]
Corn starch	Cross-linked thermoplastic starch with polyvinyl alcohol blend film significantly reduced the moisture sensitivity with improvement in tensile strength.	[48]
Leaf wood cellulose	Leaf wood cellulose fiber plasticized with wheat starch matrix composite showed improvement in transition temperature which influenced the mechanical strength.	[49]
Ulva armoricana	Green algae fiber cross-linked with polyvinyl alcohol composite showed promising mechanical properties that could be useful in the agriculture and packaging sector.	[50]
Sugar palm starch	Incorporation of agar in thermoplastic starch blend showed a slight increase in hydrophilicity, however no significant influence on density was observed.	[51]

(Table 2) cont.....

Source of Bio-Based Polymer	Application in Bioplastics Manufacturing	Reference
Musa cavendishii AAA	Thermoplastic unripe banana flour (80%) blended with metallocene-catalyzed polyethylene (20%) indicated similar mechanical properties with excellent biodegradability, compared to metallocene-catalyzed polyethylene plastic alone.	[52]
lignocellulosic material from *Eucalyptus wood*	A polymer matrix blend of low-density polyethylene and polyvinyl chloride with a different compositional ratio of lignocellulosic material demonstrated improvement in Young's modulus, however over mechanical property was reduced.	[53]

Recently a young scientist Lucy Hughes from The University of Sussex (UK), won James Dyson Award in 2019 for discovering a cost-effective, biodegradable, and compostable source of plastics originating from marine waste with red algae as a binder that could mimic the polyurethane properties [26]. The intellectual property rights of innovation were protected as MarineTex, which represents in a versatile material that can be an excellent alternative to plastic in a variety of applications. Fig. (**6**) represents the prototype of MarineTex. The fish and aquaculture processing industry generates enormous wastes that have the potential to convert into bioplastics. Approximately 40-60% of fish weight usually ends up as waste that potentially contains 20% of oil. Alone, the United Kingdom industrial fish processing plant produces 172,702 tons of fish waste annually [27]. Though fish leather is part of a folkloric tradition, especially in East Asia, it has been relived as polymers in the manufacturing of biodegradable plastics. Bioplastics manufactured using this polymer are translucent and stronger than low-density polyethylene with similar thickness to commercial plastics. In addition, bioplastics are inexpensive and degrade within four to six weeks, compared to synthetic plastics. A report estimated that Atlantic codfish produces enough waste that could be useful to fabricate 1,400 bioplastics bags [26]. Ultimately bioplastic can solve basic two problems such as the ubiquity of single-use plastics and fish waste.

Instead of using phosgene and fossil that produce toxic gas during the manufacturing of plastics, bioplastics are produced by processing fish waste through a chemical reaction using hydrogen peroxide and later reacted with amines to form a polymer similar to polyurethane [28]. The bioplastic produced using such a bio-polymer could be transformed in various sectors including cling wrap or medical wound dressings. Perhaps, vegetable and soybean oil can also create bioplastics, however, bioplastic produced using fish waste would be destined for the trash instead of growing new crops for plastics that can often be an energy-intensive process.

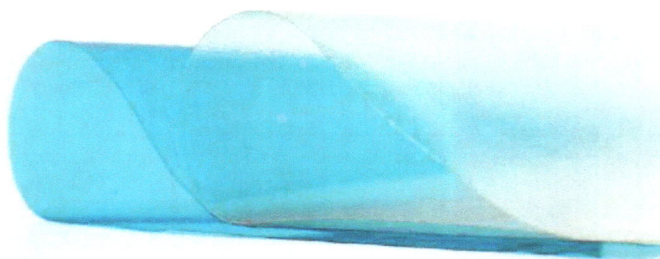

Fig. (6). MarinaTex (Dyson) is a prototype bioplastic produced using fish scale waste and red algae as a binder [26].

CONCLUSIONS AND FUTURE PROSPECTS

Exposure to plastic and plasticizer supplemented additives within synthetic plastics is omnipresent in contemporary society. The biodegradability of synthetic plastics demands instant actions due to their toxicological effects. Recycling is one of the most important possibilities to reduce the devastating effects of plastics. In addition, recycling of a wider range of packaging plastic, together with waste plastic from consumer goods such as plastic-based waste electronic and electrical equipment with obsolescence plastic vehicles accessories will further enable improvement in recovery rates of plastic waste and diversion from landfills. Furthermore, recycling also provides an opportunity to limit the consumption of oil, carbon dioxide emissions, and the quantities of waste requiring disposal. Next-generation plastic production uses nonpetroleum-based products with carbon-neutral monomers, which are non-toxic and degrade at a significant rate to prevent the accumulation of plastic debris in terrestrial and aquatic environments. These efforts could increase the process of recycling of waste plastics in an effective way to improve the environmental performance of the bio-based polymer industry.

REFERENCES

[1]   Aisbl P. Plastics the facts 2018 an analysis of European plastics production, demand and waste data. Brussels, Belgium: PlasticsEurope 2018.

[2]   Andrady AL, Neal MA. Applications and societal benefits of plastics. Philos Trans R Soc Lond B Biol Sci 2009; 364(1526): 1977-84.
[http://dx.doi.org/10.1098/rstb.2008.0304] [PMID: 19528050]

[3]   MacArthur DE, Waughray D, Stuchtey M. The new plastics economy, rethinking the future of plastics. World Economic Forum.

[4]   Thompson RC, Swan SH, Moore CJ, vom Saal FS. Our plastic age. Philos Trans R Soc Lond B Biol Sci 2009; 364(1526): 1973-6.

[http://dx.doi.org/10.1098/rstb.2009.0054] [PMID: 19528049]

[5]   Europe P. Plastics the facts, An analysis of European plastics production, demand and waste data, Brussels, Belgium. 2017.

[6]   Lebreton L, Andrady A. Future scenarios of global plastic waste generation and disposal. Palgrave Commun 2019; 5(1): 6.
[http://dx.doi.org/10.1057/s41599-018-0212-7]

[7]   Forum WE. The new plastics economy. Rethinking the future of plastics: Geneva, Switzerland 2016. Available from: https://www.weforum.org/reports/the-new-plastics-economy-rethinking-the-future-of-plastics

[8]   British Plastics Federation Oil consumption. British Plastics Federation 2019. Available from: https://www.bpf.co.uk/press/Oil_Consumption

[9]   Schmidt C, Krauth T, Wagner S. Export of plastic debris by rivers into the sea. Environ Sci Technol 2017; 51(21): 12246-53.
[http://dx.doi.org/10.1021/acs.est.7b02368] [PMID: 29019247]

[10]   Mrowiec B. Plastics in the circular economy (CE). Environmental Protection and Natural Resources 2018; 29(4): 16-9.
[http://dx.doi.org/10.2478/oszn-2018-0017]

[11]   Hoffmann RE, Koesoemo RS. Practical E&P waste minimization opportunities: moving up the 5R hierarchy. Proceedings of SPE Asia Pacific Oil and Gas Conference and Exhibition.

[12]   Europe P. Plastics the facts 2020 Plastics Europe: accelerating sustainable solutions valued by society. Brussels, Belgium 2020. Available from: https://www.plasticseurope.org/en/resources/publications/4312-plastics-facts-2020

[13]   Meran C, Ozturk O, Yuksel M. Examination of the possibility of recycling and utilizing recycled polyethylene and polypropylene. Mater Des 2008; 29(3): 701-5.
[http://dx.doi.org/10.1016/j.matdes.2007.02.007]

[14]   Koller M, Salerno A, Reiterer A, Malli H, Kettl K-H, Narodoslawsky M, *et al.* Sugarcane as feedstock for biomediated polymer production. In: Joao FG, Kaue DC, Eds. Sugarcane: production, cultivation and uses (agriculture issues and polices). Nova Science Publisher 2012; pp. 105-36.

[15]   Fabunmi O, Tabil L, Panigrahi S, Chang P. Developing Biodegradable Plastics from starch. American Soc Agri Biol Eng 2007.

[16]   Yusof FM, Wahab NA, Abdul Rahman NL, Kalam A, Jumahat A, Mat Taib CF. Properties of treated bamboo fiber reinforced tapioca starch biodegradable composite. Mater Today Proc 2019; 16: 2367-73.
[http://dx.doi.org/10.1016/j.matpr.2019.06.140]

[17]   Travalini AP, Lamsal B, Magalhães WLE, Demiate IM. Cassava starch films reinforced with lignocellulose nanofibers from cassava bagasse. Int J Biol Macromol 2019; 139: 1151-61.
[http://dx.doi.org/10.1016/j.ijbiomac.2019.08.115] [PMID: 31419552]

[18]   Florencia V, López OV, García MA. Exploitation of by-products from cassava and ahipa starch extraction as filler of thermoplastic corn starch. Compos, Part B Eng 2020; 182: 107653.
[http://dx.doi.org/10.1016/j.compositesb.2019.107653]

[19]   Engel JB, Ambrosi A, Tessaro IC. Development of biodegradable starch-based foams incorporated with grape stalks for food packaging. Carbohydr Polym 2019; 225: 115234.
[http://dx.doi.org/10.1016/j.carbpol.2019.115234] [PMID: 31521283]

[20]   Jumaidin R, Saidi ZAS, Ilyas RA, Ahmad MN, Wahid MK, Yaakob MY, *et al.* Characteristics of cogon grass fibre reinforced thermoplastic cassava starch biocomposite: water absorption and physical properties. J Adv Res Fluid Mechan Therm Sci 2019; 62(1): 43-52.

[21]   Kaisangsri N, Kerdchoechuen O, Laohakunjit N. Biodegradable foam tray from cassava starch

blended with natural fiber and chitosan. Ind Crops Prod 2012; 37(1): 542-6.
[http://dx.doi.org/10.1016/j.indcrop.2011.07.034]

[22]   Zerroukhi A, Jeanmaire T, Raveyre C, Ainser A. Synthesis and characterization of hydrophobically modified starch by ring opening polymerization using imidazole as catalyst. Stärke 2012; 64(8): 613-20.
[http://dx.doi.org/10.1002/star.201100154]

[23]   Lu DR, Xiao CM, Xu SJ. Starch-based completely biodegradable polymer materials. Express Polym Lett 2009; 3(6): 366-75.
[http://dx.doi.org/10.3144/expresspolymlett.2009.46]

[24]   Tsang YF, Kumar V, Samadar P, *et al.* Production of bioplastic through food waste valorization. Environ Int 2019; 127: 625-44.
[http://dx.doi.org/10.1016/j.envint.2019.03.076] [PMID: 30991219]

[25]   Diyana Z, Jumaidin R, Selamat MZ, Ghazali I, Julmohammad N, Huda N, *et al.* Physical Properties of Thermoplastic Starch Derived from Natural Resources and Its Blends: A Review Polym. 2021; 13(9): 1396.
[http://dx.doi.org/10.3390/polym13091396]

[26]   Retrieved from MarinaTex website. 2019. Available from: https://www.marinatex.co.uk/

[27]   Mathiesen ÁM. The state of world fisheries and aquaculture. Food and Agriculture Organization of the United Nations 2015. Available from: https://www.fao.org/3/i2727e/i2727e00.htm

[28]   Jafari H, Lista A, Siekapen MM, *et al.* Fish collagen: extraction, characterization, and applications for biomaterials engineering. Polymers (Basel) 2020; 12(10): 2230.
[http://dx.doi.org/10.3390/polym12102230] [PMID: 32998331]

[29]   Scognamiglio F, Gattia DM, Roselli G, Persia F, De Angelis U, Santulli C. Thermoplastic starch (tps) films added with mucilage from opuntia ficus indica: mechanical, microstructural and thermal characterization. Materials (Basel) 2020; 13(4): 1000.
[http://dx.doi.org/10.3390/ma13041000] [PMID: 32102225]

[30]   Nájera-García A, López-Hernández R, Lucho-Constantino C, Vázquez-Rodríguez G. Towards drylands biorefineries: valorisation of forage opuntia for the production of edible coatings. Sustainability (Basel) 2018; 10(6): 1878.
[http://dx.doi.org/10.3390/su10061878]

[31]   Chandra Mohan C, Harini K, Vajiha Aafrin B, *et al.* Extraction and characterization of polysaccharides from tamarind seeds, rice mill residue, okra waste and sugarcane bagasse for its Bio-thermoplastic properties. Carbohydr Polym 2018; 186: 394-401.
[http://dx.doi.org/10.1016/j.carbpol.2018.01.057] [PMID: 29456002]

[32]   Chandra Mohan C, Harini K, Karthikeyan S, Sudharsan K, Sukumar M. Effect of film constituents and different processing conditions on the properties of starch based thermoplastic films. Int J Biol Macromol 2018; 120(Pt B): 2007-16.
[http://dx.doi.org/10.1016/j.ijbiomac.2018.09.161] [PMID: 30267826]

[33]   Ashiuchi M, Oike S, Hakuba H, Shibatani S, Oka N, Wakamatsu T. Rapid purification and plasticization of d-glutamate-containing poly-γ-glutamate from Japanese fermented soybean food natto. J Pharm Biomed Anal 2015; 116: 90-3.
[http://dx.doi.org/10.1016/j.jpba.2015.01.031] [PMID: 25669727]

[34]   Lopes J, Gonçalves I, Nunes C, *et al.* Potato peel phenolics as additives for developing active starch-based films with potential to pack smoked fish fillets. Food Packag Shelf Life 2021; 28: 100644.
[http://dx.doi.org/10.1016/j.fpsl.2021.100644]

[35]   Suzuki S, Hikita H, Hernandez SC, Wada N, Takahashi K. Direct conversion of sugarcane bagasse into an injection-moldable cellulose-based thermoplastic via homogeneous esterification with mixed acyl groups. ACS Sustain Chem& Eng 2021; 9(17): 5933-41.

[http://dx.doi.org/10.1021/acssuschemeng.1c00306]

[36]   Gautam S, Sharma B, Jain P. Dynamic shear rheological study of soy protein isolate/poly(vinyl alcohol) bionanocomposites reinforced with montmorillonite nanoparticles. Polym Compos 2021; 42(5): 2349-59.
[http://dx.doi.org/10.1002/pc.25982]

[37]   Deralia PK, du Poset AM, Lund A, Larsson A, Ström A, Westman G. Oxidation level and glycidyl ether structure determine thermal processability and thermomechanical properties of arabinoxylan-derived thermoplastics. ACS Appl Bio Mater 2021; 4(4): 3133-44.
[http://dx.doi.org/10.1021/acsabm.0c01550] [PMID: 35014401]

[38]   Yang Y, Zhou H, Xiao Y, *et al.* Hydrophobic thermoplastic starch supramolecularly-induced by a functional sucrose based ionic liquid crystal. Carbohydr Polym 2021; 255: 117363.
[http://dx.doi.org/10.1016/j.carbpol.2020.117363] [PMID: 33436196]

[39]   Xu J, Sagnelli D, Faisal M, *et al.* Amylose/cellulose nanofiber composites for all-natural, fully biodegradable and flexible bioplastics. Carbohydr Polym 2021; 253: 117277.
[http://dx.doi.org/10.1016/j.carbpol.2020.117277] [PMID: 33278948]

[40]   Santos TA, Spinacé MAS. Sandwich panel biocomposite of thermoplastic corn starch and bacterial cellulose. Int J Biol Macromol 2021; 167: 358-68.
[http://dx.doi.org/10.1016/j.ijbiomac.2020.11.156] [PMID: 33278430]

[41]   González K, Iturriaga L, González A, Eceiza A, Gabilondo N. Improving mechanical and barrier properties of thermoplastic starch and polysaccharide nanocrystals nanocomposites. Eur Polym J 2020; 123: 109415.
[http://dx.doi.org/10.1016/j.eurpolymj.2019.109415]

[42]   Hassan MM, Le Guen MJ, Tucker N, Parker K. Thermo-mechanical, morphological and water absorption properties of thermoplastic starch/cellulose composite foams reinforced with PLA. Cellulose 2019; 26(7): 4463-78.
[http://dx.doi.org/10.1007/s10570-019-02393-1]

[43]   Paiva D, Pereira A, Pires A, Martins J, Carvalho L, Magalhães F. Reinforcement of thermoplastic corn starch with crosslinked starch/chitosan microparticles. Polymers (Basel) 2018; 10(9): 985.
[http://dx.doi.org/10.3390/polym10090985] [PMID: 30960910]

[44]   Fazeli M, Keley M, Biazar E. Preparation and characterization of starch-based composite films reinforced by cellulose nanofibers. Int J Biol Macromol 2018; 116: 272-80.
[http://dx.doi.org/10.1016/j.ijbiomac.2018.04.186] [PMID: 29729338]

[45]   González K, Retegi A, González A, Eceiza A, Gabilondo N. Starch and cellulose nanocrystals together into thermoplastic starch bionanocomposites. Carbohydr Polym 2015; 117: 83-90.
[http://dx.doi.org/10.1016/j.carbpol.2014.09.055] [PMID: 25498612]

[46]   Niazi MBK, Zijlstra M, Broekhuis AA. Understanding the role of plasticisers in spray-dried starch. Carbohydr Polym 2013; 97(2): 571-80.
[http://dx.doi.org/10.1016/j.carbpol.2013.04.074] [PMID: 23911487]

[47]   Moreira FKV, Marconcini JM, Mattoso LHC. Solid state ball milling as a green strategy to improve the dispersion of cellulose nanowhiskers in starch-based thermoplastic matrices. Cellulose 2012; 19(6): 2049-56.
[http://dx.doi.org/10.1007/s10570-012-9768-3]

[48]   Liu Z, Jiang M, Bai X, Dong X, Tong J, Zhou J. Effect of postcrosslinking modification with glutaraldehyde on the properties of thermoplastic starch/poly(vinyl alcohol) blend films. J Appl Polym Sci 2012; 124(5): 3774-81.
[http://dx.doi.org/10.1002/app.35382]

[49]   Avérous L, Fringant C, Moro L. Plasticized starch–cellulose interactions in polysaccharide composites. Polymer (Guildf) 2001; 42(15): 6565-72.

[http://dx.doi.org/10.1016/S0032-3861(01)00125-2]

[50]   Chiellini E, Cinelli P, Ilieva VI, Martera M. Biodegradable thermoplastic composites based on polyvinyl alcohol and algae. Biomacromolecules 2008; 9(3): 1007-13.
[http://dx.doi.org/10.1021/bm701041e] [PMID: 18257530]

[51]   Jumaidin R, Sapuan S, Jawaid M, Ishak M, Sahari J. Effect of agar on physical properties of thermoplastic starch derived from sugar palm tree. Pertanika J Sci Technol 2017; 25: 1235-48.

[52]   Martín Martínez ES, Aguilar Méndez MA, Sánchez Solís A, Vieyra H. Thermoplastic biodegradable material elaborated from unripe banana flour reinforced with metallocene catalyzed polyethylene. Polym Eng Sci 2015; 55(4): 866-76.
[http://dx.doi.org/10.1002/pen.23954]

[53]   Georgopoulos ST, Tarantili PA, Avgerinos E, Andreopoulos AG, Koukios EG. Thermoplastic polymers reinforced with fibrous agricultural residues. Polym Degrad Stabil 2005; 90(2): 303-12.
[http://dx.doi.org/10.1016/j.polymdegradstab.2005.02.020]

CHAPTER 8

Potential Application of Biopolymers in the Textile Industry

Abstract: Textile configurations are derived from two major sources such as ancient handicraft and modern scientific inventions. Textile fabrication using polymeric fiber is one of the fastest-growing sectors since the 19th century and is currently the second-largest manufacturing industry after information technology. Although polymers are predominantly used in the development of dosage forms, however recent devolment in natural polymer chemistry reflects its association with the production of plastics, fibers, elastomers, *etc*. Innovation using natural polymer fibers-based textile could serve as an alternative capable of replacing synthetic polymer-based fibers. Polymers, especially fibers contribute significantly to the manufacturing of textiles. Moreover, copolymerization of fabrics fibers with excipients demonstrated potential for the development of materials useful in various biomedical applications. Furthermore, to understand the fundamental characteristics of polymeric fibers including structural composition, morphological features such as crystallinity, and orientation, a comprehensive skill is necessary. This chapter discusses the basic materials used in the fabrication of textile products, with emphasizes on bio-based polymers as an alternative to synthetic polymers in the production of fabrics.

Keywords: Bio-based polymer, Chitosan, Fabrics, Fibers, Textile.

INTRODUCTION

The invention of rayon in 1903 prompted a revolution that led to the development of synthetic fiber in 1935 [1]. The use of fiber or yarns in the manufacturing of fabrics is one of the unique creations of human history. Today fiber industry has evolved into a complex field with rising interest among researchers. The word textile is derived from the Latin word "*textilis*" and a French word "*texere*", denoting "*to weave*" refed to woven fabric [2]. Textile materials are fabricated by the interweaving of either yarns or fibers (Fig. **1**). Textile Exchange's new preferred fiber and materials market report reveals that the global fiber production has doubled in the last 20 years, reaching an all-time high of 111 million metric tons and could potentially progress to 146 million metric tons by 2030 [3]. The annual global fabric market and production in percentage are presented in Figs. (**2** and **3**). The fibers are manufactured through the process of interlacing such as twisting, knitting, spinning, *etc*. Twisting or weaving includes basic weaves, twill, and stain while knitting includes weft and wrap process. Recent reports indicate that knitting is rapidly gaining interest among textile manufacturers [4].

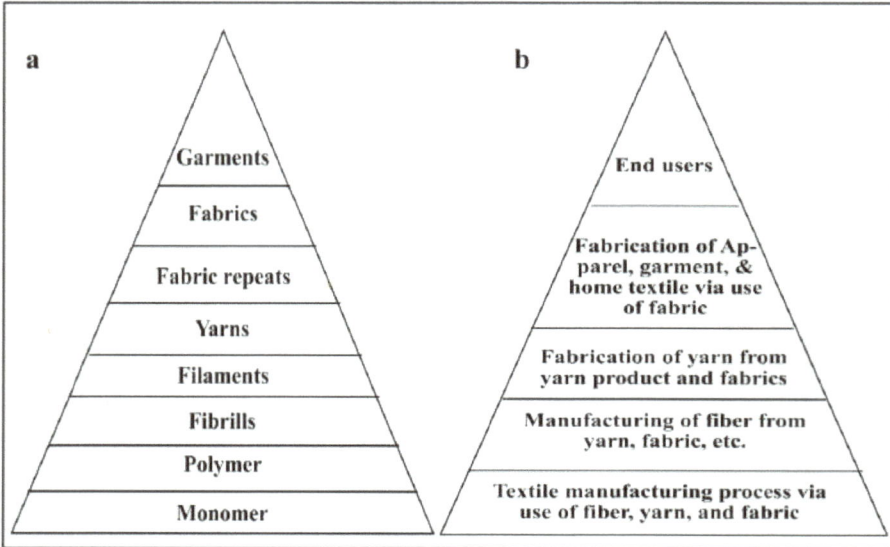

Fig. (1). The hierarchy of textiles (a), textile manufacturing process from fiber to fabrics (b).

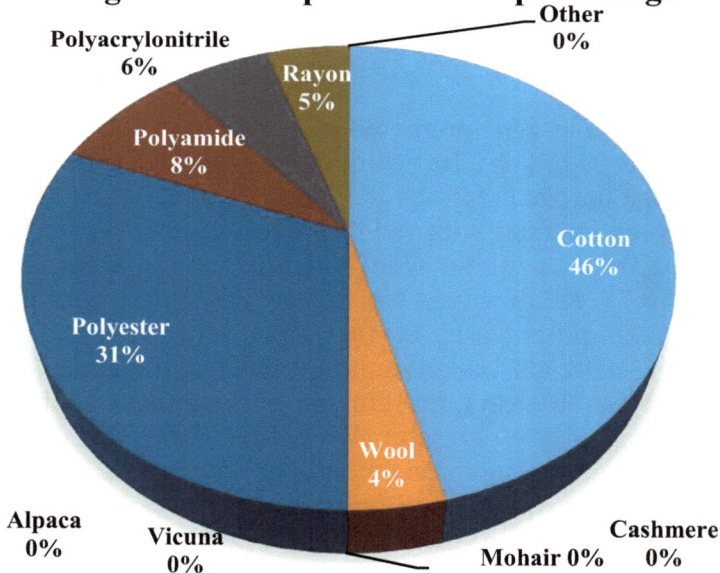

Fig. (2). The global market of fiber in pounds (for interpretation of the results to color in this Fig. legend, the reader is referred to either web version of this chapter or color print).

Textile fibers are symbolized as essential components of modern society that provide a physical structure for human comfort with sustainability. Textile fabrics are used in clothing, bedding, home furnishings, liquid filtrations, insulation materials in buildings, cleaning wipes, dental brush, wound bandages, sutures, *etc*. Currently, textile fabrics' manufacturer principally relies on synthetic petroleum-based fibers, which adversely impact the environment through global warming and pollution [5]. The rise in demand, as well as diminution in petroleum oil reserves with fluctuation in the international market value, indicates a transition from petroleum-based resources to bio-based polymers. Bio-based polymers are gaining interest in textile industries and are considered as a green alternative to synthetic petroleum fibers. Moreover, the challenges are not only associated with the pricing, but also with the derisory properties of the synthetic polymers for the textile industry.

Global fiber production (million mt)

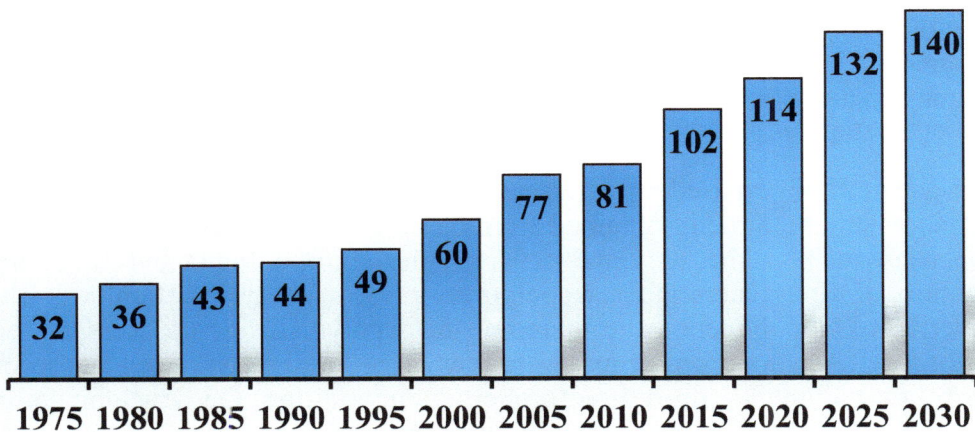

Fig. (3). Global fiber production in million metric ton projection until 2030.

Biopolymers in the Textile Industry

The earliest evidence of weaving, closely related to basketry, dates from the Neolithic culture of about 5000 BC [6]. Silk, cotton, and wool are among the natural fibers used in the manufacturing of fabrics for thousands of years, around 3000 BC in India [7]. However, due to their inherent deficiencies, applications are limited to raw form. Bernigaud de Chardonnet in 1884 invented and patented synthetic fiber also known as artificial silk, which was the first manmade fiber [8].

Developments in synthetic fiber brought a revolution in fabric material appearance that shows similar to multiple uses of plastics. Subsequent developments promoting the understanding of the macromolecular structure of a polymer by Hermann Standinger improved the invention of fiber. Moreover, DuPont de Nemour's discovery of nylon in 1938 brought a significant reform into the textile manufacturing economy, with a technology transfer in military protective apparel [9]. In addition, in 1970, Dupont successfully developed a high-strength fiber aramid with poly (p-phenylene terephthalamide), ultrahigh molecular weight polyethylene, and carbon fiber [10].

Though there are several methods for the production of polymeric fibers, melt spinning, wet spinning, and dry spinning are frequently used in the manufacturing process [11]. Dry spinning and wet spinning are used for potential high and low volatile polymers, respectively. Whereas, melt spinning is preferred for crystalline polymers that melt during the spinning process. A surge in the demand for fiber products prompted researchers to develop various other commercial methods in the production of fabrics such as gel spinning, electrostatic spinning, and integrated composite spinning [12]. High mechanical strength fabrics are produced by Gel spinning techniques, while the nanometric diameter of multi-functional fabrics is produced using electrostatic spinning technology. Moreover, combined filaments yarns with staple fibers are produced by the integrated spinning method, which produces fiber at a considerable high speed compared with other staple processes.

First-generation bio-based polymers such as starch and carbohydrate-related raw excipients can be directly obtained from natural resources. However, undesired properties such as the presence of impurities, morphology, and solid states structure of first-generation raw polymers present several difficulties. These properties of biopolymer can be altered by the depolymerization process to yield a monomer. The monomers can be further used in the synthesis of second-generation polymers with desired properties. In addition, the development of a biopolymer from industrial, agricultural, marine, and household waste is also gaining importance in the textile industry.

Fibers originating through the natural process of plants or animals are generally degradable concerning time and environment. Whereas synthetic fibers developed through chemically synthesized materials or polymers may require some time, compared with natural products (Fig. **4**). Synthetic polymeric fibers commercially used in the fabrication of fabrics including polyamide-nylon, polypropylene, polytetrafluoroethylene, polyethylene terephthalate, polybutylene terephthalate, polyester, phenol-formaldehyde, polyvinyl alcohol, polyvinyl chloride, polyolefin, polyethylene, low-density polyethylene, high-density polyethylene, and ultrahigh molecular weight polyethylene [13] are non-degradable or take years to degrade. However, natural bio-based polymers such as cellulose, polypeptides, chitosan, silk, cotton, jute, keratin, wool, fibroin [14], *etc.*, have shown potential application in the fabrication of textile fabrics and biomedical applications that degrade in a short duration of time with no or less impact on the environment.

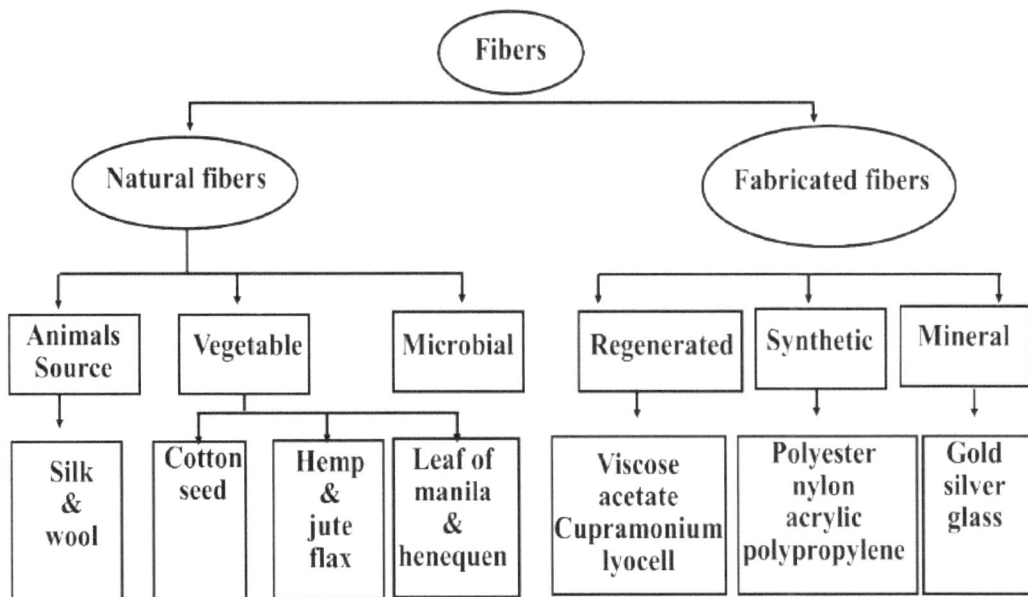

Fig. (4). Classification of fibers.

Silk, derived from the silkworm (*Bombyx mori*) is composed of majorly two proteins, fibroin (fibrous protein) and sericin (globular protein) [15]. Fibroin has been used in textile manufacturing since ancient times in the fabrication of several biomaterials and fabrics, whereas sericin is considered a waste product from the silk waste industry consisting of variable amino acids with potential pharmaceutical and biomedical applications [16] (Fig. **5**). Sericin's affirmative effect on keratinocytes and fibroblast directs its potential application in the development of biomaterials that could be useful for skin and bone tissue engineering [17]. The thin, long, light, soft, and crystalline silk fiber is a valuable material used in the production of various categories of textiles including apparel, sarees, suiting's, curtains, luxurious interiors, biodegradable sutures, *etc.* Moreover, silk proteins are biodegradable materials with numerous reactive functional groups that provide possibilities of co-polymerization with other polymers to be used in novel drug deliveries. Silk fibroin forms a β-sheet structure on the polymeric chain that could produce an excellent tensile strength ranging between 0.171 – 0.468 MPa [18]. In addition, silk is also known for its multifarious properties such as absorbency, dyeing affinity, thermo-tolerance, insulation, and luster [19]. However, due to several limitations such as reduction in mechanical strength in the presence of sunlight, chlorine bleach, perspiration, absorbing body oils, formation of water spots, and color fading associated with the use of silk alone, it is reportedly crosslinking with other polymers in fabrics production. Though the utility of silk alone in textiles indicates several limitations, the global capital market is growing significantly with other cross-linked polymers. The statistical data represented that the current size and share of the silk market are USD 21.45 billion by revenue and it is expected to rise at a rate of 9.5% to reach a total of 28.71 billion by 2026 [20] (Fig. **6**).

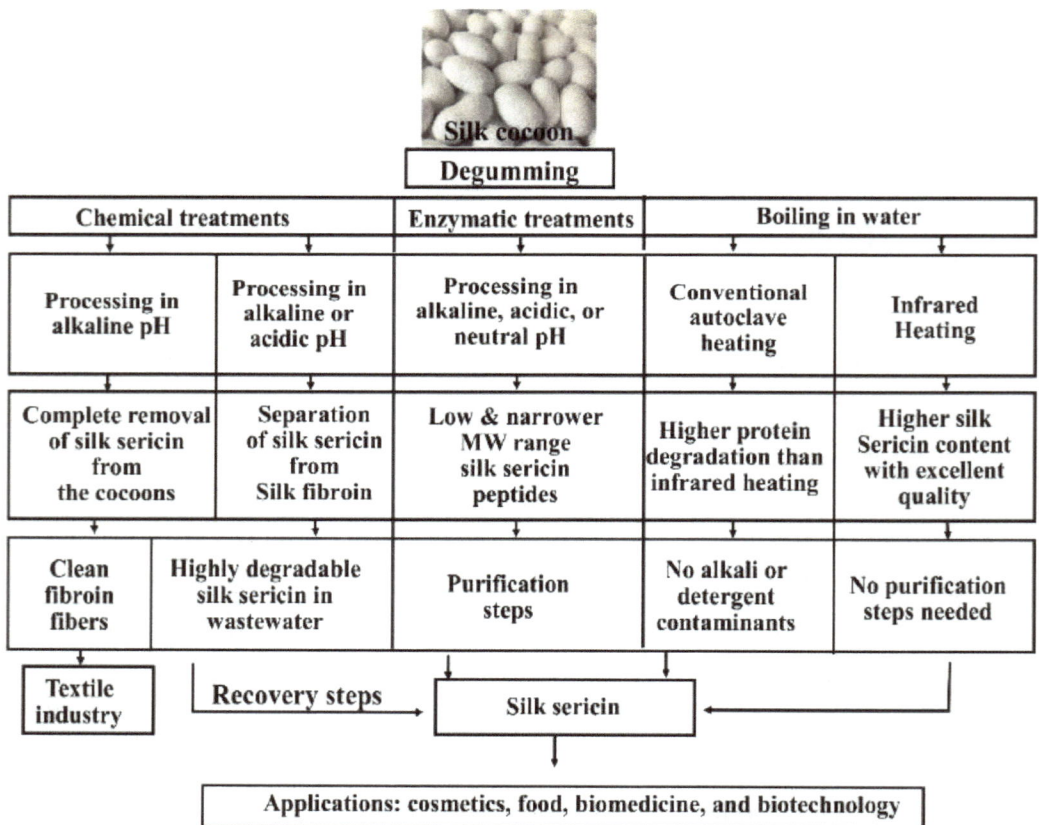

Chemical treatments		Enzymatic treatments	Boiling in water	
Processing in alkaline pH	Processing in alkaline or acidic pH	Processing in alkaline, acidic, or neutral pH	Conventional autoclave heating	Infrared Heating
Complete removal of silk sericin from the cocoons	Separation of silk sericin from Silk fibroin	Low & narrower MW range silk sericin peptides	Higher protein degradation than infrared heating	Higher silk Sericin content with excellent quality
Clean fibroin fibers	Highly degradable silk sericin in wastewater	Purification steps	No alkali or detergent contaminants	No purification steps needed

Textile industry ← Recovery steps ← Silk sericin

Applications: cosmetics, food, biomedicine, and biotechnology

Fig. (5). Production and purification of silk sericin from the silk cocoon.

Natural textile fibers significantly dominated the global textile market until 1950. The Food and Agriculture Organization reported that the use of cellulose and synthetic polymers with traditional fibers unexpectedly changed the textile fibers consumption by 52% during 1948-58 [21]. Though natural fibers alone almost lost the stake in global textile fiber fabrication, their production and consumption are slowly rising *via* the use of polymerized various polymers and chemicals. Chitosan is a unique and the second most abundant natural bio-based polymer extracted from various resources including fungi, algae, echiruda, annelida, mollusca, cnidaria, achelminthes, entroprocta, bryozoa, phoronida, brachiopoda, arthropoda, and ponogophora [22]. However, researchers commonly report the various applications of the deacetylated form of chitin originating from shrimp, crab, lobster, crayfish scale, and oyster [23]. Chitin occurs in three polymeric forms based on the differing degree of hydration, size of the unit cell, chitin chain per unit cell as α, β, and γ chitin [24]. Chitosan is well-known for its anti-bacterial and film-forming properties with good mechanical and thermal stability in various applications such as food packaging [25, 26], scaffold for dermal applications [27], matrix release retardant [28], *etc.*

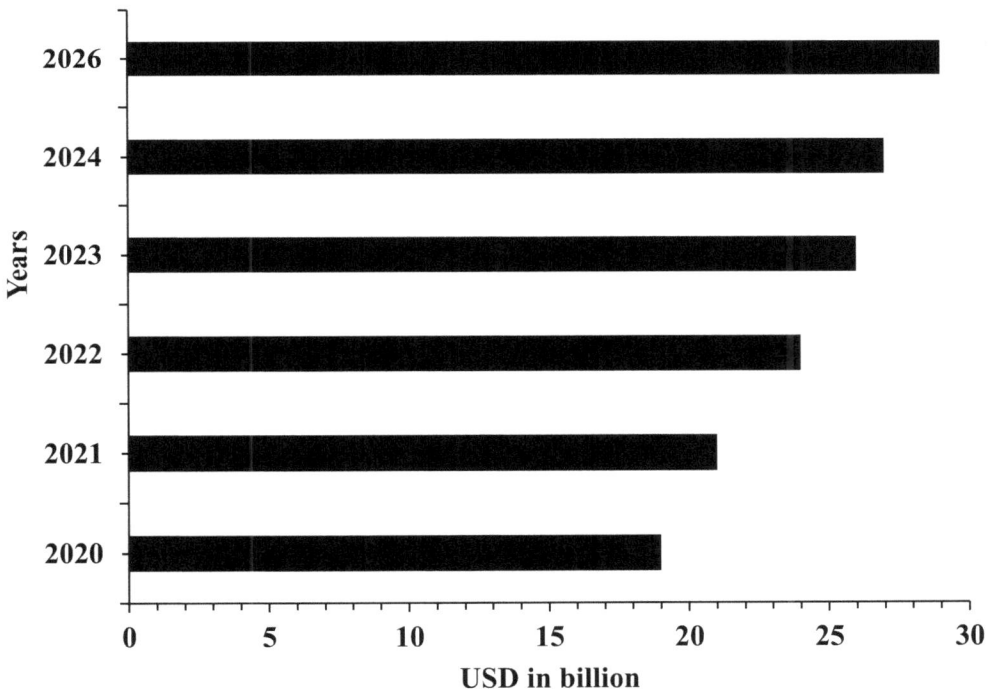

Fig. (6). Silk market size, trends, and forecast between 2020 – 2026 by revenue (for interpretation of the results to color in this Fig legend, the reader referred to either web version of this chapter or color print).

Chitosan is a polymer chosen for the development of functional fabrics due to its excellent biocompatibility, biodegradability, and ecological safety. Moreover, copolymerization of chitosan with fiber and yarn has demonstrated excellent anti-wrinkle, antistatic, and dyeing properties as finishing agents or modifiers for textiles [29]. In addition, textile innovators have also developed advanced products in combination of fibers and chitosan for multifunctional applications including antibacterial chitosan nanofibers [30, 31], enhanced mechanical strength of surgical materials [32, 33], dye degradation [34], fire extinguishers polyester-cotton fabric [35], *etc.*

Alginate is a water-soluble naturally occurring biopolymer polysaccharide extracted from bacteria and brown algae cell walls of various species including *laminaria, Macrocystis,* and *Ascophyllum* [36]. Alginate and its derivatives exist as linear unbranched polysaccharides consisting of – mannuronic and – guluronic acid units organized in an asymmetrical pattern. The presence of free hydroxyl and carboxyl groups along the backbone makes alginate an ideal candidate for the functionalization and improvement of fibers with desired properties. Various alginate salt fibers have been manufactured through the spinning process for biomedical applications considering the excellent water holding and gel-forming properties. Moreover, recent studies indicated that alginate-based fibers have been successfully developed as scaffolds in tissue engineering [37]. The recent potential application of biopolymers in the textile industry is presented in Table **1**.

Table 1. Potential industrial textile application of bio-based polymers.

Polysaccharides	Application	Reference
Carthamus tinctorius	Cross-linked bioactive bio-based polymer-alginate complex revitalized with textile material for potential diabetic foot wound healing.	[38]
Xylans derived from beech wood and oat spelts	Polyethylene terephthalate fabric functionalized with cationic xylans demonstrated excellent antimicrobial efficacy against *Staphylococcus aureus* and *Escherichia coli.*	[39]
Commercial chitosan	Chitosan-poly(acrylic acid) spun biodegradable fiber with excellent softness and toughness.	[40]
Commercial modified cellulose	Microcrystalline cellulose and zinc chloride hydrate mono and multi-filament alcogel spun microfibers for potential application in woundcare.	[41]
Opuntia megacantha	Activated carbon from polysaccharides of *Opuntia megacantha* with a surface between 500 to 1000 m^2/g demonstrated excellent adsorption of textile liquid waste.	[42]
Corchorus Olitorius	Glycine treated jute yarn demonstrated improvement in thermomechanical properties, helpful in the development of yarn-based woven or multi-axial textile architecture.	[43]
Commercial chitosan	Amendment of cadmium polluted soils with textile waste biochar coated with natural polymer lowered cadmium bioavailability and reduced associated environmental and human health risks.	[44]

(Table 1) cont.....

Polysaccharides	Application	Reference
Commercial chitosan	Fabric treated with chitosan enhanced the drying efficiency and improved wash fastness.	[45]
Cationic guar, crosslinked hydroxypropyl guar	Improvement in fabric conditioner performance with the deposition of surfactant and polysaccharides.	[46]
Guar gum	Thin polymeric membrane coated guar gum-silver activated polyester fabric demonstrated promising potential in removing organic contaminants from the effluent of printing, dyeing, and detergent industries.	[47]
Potato starch and polyvinyl alcohol	Ecofriendly disposable nonwoven starch-polyvinyl alcohol ultrafine fibrous membrane showed improvement in mechanical properties.	[48]
Gellan gum	Electrospun nanofibers based on polysaccharides demonstrated a consolidated approach in tissue engineering and regenerative medicine.	[49]
Guar gum	Guar gum-grafted polyacrylamide cross-linked with borax based adsorbent in the removal of industrial dye-RB-19	[50]
Lignin and chitosan	Natural filler-based flexible polyurethane foam thermosets demonstrated commercial application in the fabrication of mattresses, car seats, and insulating materials in textile as well as cosmetic industries.	[32]
***Carica papaya* seeds with chitosan**	Activated carbon-chitosan-*Carica papaya* composite with reuse potential demonstrated efficient adsorbent dye removal from wastewater.	[51]
Chitosan-polyvinyl alcohol	Electrospinning wound dressing developed with recombinant human erythropoietin/aloe-vera gel releasing polyvinyl-alcohol/chitosan significant wound closure, compared with control.	[52]
Chitosan	Cotton fabric treated with different concentrations of chitosan and silver nanoparticles showed resistance towards microbes and ultraviolet protection factors with a reduction in air permeability and water absorbance.	[53]
Potato and wheat starch	Amylase dispersion demonstrated a green approach in the removal of aged starch-based glues from textiles.	[54]
Commercial chitosan	Pea nut waste shell reduced silver nanoparticles with chitosan-coated cellulosic fabric resulted in synergistic antioxidant and antimicrobial efficacy.	[55]
Aloe vera cactus leaves	Cellulose nanocrystals isolated from leaves demonstrated promising potential towards the fabrication of textile and packaging materials.	[56]
***Sargassum wightii* and *padina tetrastromatica* seaweeds**	Sodium alginate isolated from *Sargassum wightii* coated textile fabric showed higher wound healing efficacy compared to sodium alginate isolated from *padina tetrastromatica*.	[57]

(Table 1) cont.....

Polysaccharides	Application	Reference
Cochlospermum religiosum	Natural gum grafted with poly(acrylamide co-poly-*N*-methyl acrylamide) hydrogel demonstrated potential dye adsorbing capacity to treat the textile wastewater.	[58]
Xanthan and psyllium gum	Gum xanthan-psyllium hybrid backbone graft copolymerized with polyacrylic acid-copolyitaconic acid chains based adsorbent for the treatment of dye contaminated textile water.	[59]
Psyllium, alginate, and chitosan	Psyllium containing alginate chitosan fibers indicated excellent tensile property with potential wound dressing application.	[60]
Corn starch	Amylopectin and hydroxyethyl starch grafted acrylamide showed excellent flocculating efficacy in a paper mill and textile dye effluent.	[61]
Chitosan	Chitosan-coated perlite beads showed promising efficacy in the removal of dye from wastewater.	[62]
Potato starch and alginate	Starch, alginate with micro-fibrous mineral clay-based nanocomposite foam showed good mechanical properties with fire resistance as eco-friendly insulating materials.	[63]
Alginate	Alginate-montmoril-ionite composite bead showed the improved surface area with dye adsorption capacity.	[64]

CONCLUSIONS AND FUTURE PROSPECTS

Exploitation and advent of bio-based polymers in various textile manufacturing processes have gained remarkable consideration and have the potential to replace the commonly used synthetic polymers in the future. Global warming and diminution of resources due to the fabrication of textile using synthetic materials have increased attention towards sustainable development. The replacement of conservative synthetic resources with green alternative technologies in the textile industry has proven several benefits. The copolymerization of fibers or yarn with bio-based polymers is slowly reducing the dependency of the textile industry on petroleum resources meanwhile significantly reducing the detrimental effects on the environment. In addition, the rise in the utilization of bio-based polymers by the textile industry not only contributes to the fabrication of eco-friendly fabrics but also produces low-weight fabrics, with a minimum cost of production. Despite a radical shift in the use of synthetic polymer to bio-based materials in the fabrication of fabrics, the process is slow due to several drawbacks with natural polymers that impede the effective use of bio-based polymers in the textile industry.

REFERENCES

[1]   Keist CN. Rayon and its impact on the fashion industry at its introduction. MS dissertation Ann Arbor: Iowa State University 2009.

[2]   Arfin T. Reactive and functional polymers. In: Shahid ul-Isslam, Butola BS, Eds. Advanced functional textiles and polymers. Wiley, Scrivener Publishing LLC 2019; pp. 291-308.

[3]   Pepper LR. Preferred Fiber and Materials Market Report (PFMR) Released. 2020. Available from: https://textileexchange.org/2020-preferred-fiber-and-materials-market-report-pfmr-released-2/

[4]   Leong KH, Ramakrishna S, Huang ZM, Bibo GA. The potential of knitting for engineering composites—a review. Compos, Part A Appl Sci Manuf 2000; 31(3): 197-220.
[http://dx.doi.org/10.1016/S1359-835X(99)00067-6]

[5]   Palacios-Mateo C, van der Meer Y, Seide G. Analysis of the polyester clothing value chain to identify key intervention points for sustainability. Environ Sci Eur 2021; 33(1): 2.
[http://dx.doi.org/10.1186/s12302-020-00447-x] [PMID: 33432280]

[6]   Whewell CSaA. Textile. Encyclopedia Britannica 2020. Available from: https://www.britannica.com/topic/textile

[7]   Ruchi Shukla MN. Cotton Textile Industry in India 2019. Available from: https://www.investindia.gov.in/team-india-blogs/cotton-textile-industry-india

[8]   Britannica TEoE. Hilaire Bernigaud, count de Chardonnet 2021. Available from: https://www.britannica.com/biography/Louis-Marie-Hilaire-Bernigaud-comte-de-Chardonnet

[9]   Anderson KJ. Fifty Years of Nylon. MRS Bull 1988; 13(5): 38-9.
[http://dx.doi.org/10.1557/S0883769400065672]

[10]   Rao Y, Waddon AJ, Farris RJ. Structure–property relation in poly(p-phenylene terephthalamide) (PPTA) fibers. Polymer (Guildf) 2001; 42(13): 5937-46.
[http://dx.doi.org/10.1016/S0032-3861(00)00905-8]

[11]   Whewell CSaA. Textile. ncyclopedia Britannica 2020. Available from: https://www.britannica.com/topic/textile

[12]   Kuo CJ, Lan WL. Gel spinning of synthetic polymer fibres. In: Zhang D, Ed. Advances in filament yarn spinning of textiles and polymers. Woodhead Publishing 2014; pp. 100-12.
[http://dx.doi.org/10.1533/9780857099174.2.100]

[13]   Goldade VA, Vinidiktova NS. Novel crazing technology applications. In: Goldade VA, Vinidiktova NS, Eds. Crazing technology for polyester fibers. Woodhead Publishing 2017; pp. 139-59.
[http://dx.doi.org/10.1016/B978-0-08-101271-0.00007-3]

[14]   Schuster KCRC, Eichinger D, Schmidtbauer J, Aldred P, Firgo H. Environmentally friendly lyocell fibers. In: Wallenberger FT, Weston NE, Eds. Natural Fibers, plastics and composites. Boston, MA: Springer 2004; pp. 123-46.
[http://dx.doi.org/10.1007/978-1-4419-9050-1_9]

[15]   Kunz RI, Brancalhão RMC, Ribeiro LFC, Natali MRM. Ribeiro LdFC, Natali MRM. Silkworm sericin: properties and biomedical applications. BioMed Res Int 2016; 2016: 1-19.
[http://dx.doi.org/10 1155/2016/8175701] [PMID: 27965981]

[16]   Aramwit P, Siritientong T, Srichana T. Potential applications of silk sericin, a natural protein from textile industry by-products. Waste Manag Res 2012; 30(3): 217-24.
[http://dx.doi.org/10.1177/0734242X11404733] [PMID: 21558082]

[17]   Lamboni L, Li Y, Liu J, Yang G. Silk Sericin-functionalized bacterial cellulose as a potential wound-healing biomaterial. Biomacromolecules 2016; 17(9): 3076-84.
[http://dx.doi.org/10.1021/acs.biomac.6b00995] [PMID: 27467880]

[18]   Nguyen TP, Nguyen QV, Nguyen VH, *et al.* Silk fibroin-based biomaterials for biomedical applications: a review. Polymers (Basel) 2019; 11(12): 1933.
[http://dx.doi.org/10.3390/polym11121933] [PMID: 31771251]

[19]   Study on polyacrylic acid treated silk. In: Parthiban M, Srikrishnan MR, Kandhavadivu P, Eds.

Sustainability in fashion and apparels: challenges and solutions. WPI Publishing 2017; p. 10.

[20] Silk Market is Segmented By Type, Application and region - global industry analysis on size, share, growth, investment and forecast. market data forecast 2021. Available from: https://www.marketdataforecast.com/market-reports/silk-market

[21] Datta RK, Nanavaty M. Global silk industry: a complete source book. Universal Publishers 2005; pp. 294-32.

[22] Ahmed SIS. Chitosan and its derivatives: a review in recent innovations. Int J Pharm Sci Res 2015; 6(1): 14-30.

[23] Lopes PP, Tanabe EH, Bertuol DA. Chitosan as biomaterial in drug delivery and tissue engineering. In: Gopi S, Thomas S, Pius A, Eds. Handbook of chitin and chitosan. Elsevier 2020; pp. 407-31.
[http://dx.doi.org/10.1016/B978-0-12-817966-6.00013-3]

[24] Jang MK, Kong BG, Jeong YI, Lee CH, Nah JW. Physicochemical characterization of? -chitin? -chitin, and? -chitin separated from natural resources. J Polym Sci A Polym Chem 2004; 42(14): 3423-32.
[http://dx.doi.org/10.1002/pola.20176]

[25] Eze FN, Jayeoye TJ, Singh S. Fabrication of intelligent pH-sensing films with antioxidant potential for monitoring shrimp freshness *via* the fortification of chitosan matrix with broken Riceberry phenolic extract. Food Chem 2022; 366: 130574.
[http://dx.doi.org/10.1016/j.foodchem.2021.130574] [PMID: 34303209]

[26] Singh S, Nwabor OF, Syukri DM, Voravuthikunchai SP. Chitosan-poly(vinyl alcohol) intelligent films fortified with anthocyanins isolated from *Clitoria ternatea* and *Carissa carandas* for monitoring beverage freshness. Int J Biol Macromol 2021; 182: 1015-25.
[http://dx.doi.org/10.1016/j.ijbiomac.2021.04.027] [PMID: 33839180]

[27] Shirzaei Sani I, Rezaei M, Baradar Khoshfetrat A, Razzaghi D. Preparation and characterization of polycaprolactone/chitosan-g-polycaprolactone/hydroxyapatite electrospun nanocomposite scaffolds for bone tissue engineering. Int J Biol Macromol 2021; 182: 1638-49.
[http://dx.doi.org/10.1016/j.ijbiomac.2021.05.163] [PMID: 34052267]

[28] Balan P, Indrakumar J, Murali P, Korrapati PS. Bi-faceted delivery of phytochemicals through chitosan nanoparticles impregnated nanofibers for cancer therapeutics. Int J Biol Macromol 2020; 142: 201-11.
[http://dx.doi.org/10.1016/j.ijbiomac.2019.09.093] [PMID: 31604079]

[29] Croitoru C, Roata IC. Alteration and enhancing the properties of natural fibres. In: Mondal MIH, Ed. Fundamentals of natural fibres and textiles. Woodhead Publishing 2021; pp. 367-405.
[http://dx.doi.org/10.1016/B978-0-12-821483-1.00006-1]

[30] Li J, Tian X, Hua T, *et al.* Chitosan natural polymer material for improving antibacterial properties of textiles. ACS Appl Bio Mater 2021; 4(5): 4014-38.
[http://dx.doi.org/10.1021/acsabm.1c00078] [PMID: 35006820]

[31] Massella D, Argenziano M, Ferri A, *et al.* Bio-Functional textiles: combining pharmaceutical nanocarriers with fibrous materials for innovative dermatological therapies. Pharmaceutics 2019; 11(8): 403.
[http://dx.doi.org/10.3390/pharmaceutics11080403] [PMID: 31405229]

[32] Husainie SM, Khattak SU, Robinson J, Naguib HE. A comparative study on the mechanical properties of different natural fiber reinforced free-rise polyurethane foam composites. Ind Eng Chem Res 2020; 59(50): 21745-55.
[http://dx.doi.org/10.1021/acs.iecr.0c04006]

[33] Attia NF, Ebissy AAE, Morsy MS, Sadak RA, Gamal H. Influence of textile fabrics structures on thermal, uv shielding, and mechanical properties of textile fabrics coated with sustainable coating. J Nat Fibres 2020; pp. 1-8.

[34]   Preethi S, Abarna K, Nithyasri M, *et al.* Synthesis and characterization of chitosan/zinc oxide nanocomposite for antibacterial activity onto cotton fabrics and dye degradation applications. Int J Biol Macromol 2020; 164: 2779-87.
[http://dx.doi.org/10.1016/j.ijbiomac.2020.08.047] [PMID: 32777425]

[35]   Leistner M, Abu-Odeh AA, Rohmer SC, Grunlan JC. Water-based chitosan/melamine polyphosphate multilayer nanocoating that extinguishes fire on polyester-cotton fabric. Carbohydr Polym 2015; 130: 227-32.
[http://dx.doi.org/10.1016/j.carbpol.2015.05.005] [PMID: 26076621]

[36]   Pereira L, Cotas J. Alginates a general overview. In: Pereira L, Ed. Alginates recent uses of this natural polymer IntechOpen Book Series: Biochemistry. IntechOpen Ltd 2020; pp. 3-16.
[http://dx.doi.org/10.5772/intechopen.88381]

[37]   Tao F, Cheng Y, Tao H, *et al.* Carboxymethyl chitosan/sodium alginate-based micron-fibers fabricated by emulsion electrospinning for periosteal tissue engineering. Mater Des 2020; 194: 108849.
[http://dx.doi.org/10.1016/j.matdes.2020.108849]

[38]   El-Ghoul Y, Alminderej FM. Bioactive and superabsorbent cellulosic dressing grafted alginate and *Carthamus tinctorius* polysaccharide extract for the treatment of chronic wounds. Text Res J 2021; 91(3-4): 235-48.
[http://dx.doi.org/10.1177/0040517520935213]

[39]   Velkova N, Zemljic LF, Saake B, Strnad S. Adsorption of cationized xylans onto polyethylene terephthalate fabrics for antimicrobial medical textiles. Text Res J 2019; 89(4): 473-86.
[http://dx.doi.org/10.1177/0040517517748512]

[40]   Ohkawa K, Ando M, Shirakabe Y, *et al.* Preparing chitosan-poly(acrylic acid) composite fibers by self-assembly at an aqueous solution interface. Text Res J 2002; 72(2): 120-4.
[http://dx.doi.org/10.1177/004051750207200205]

[41]   Rostamitabar M, Subrahmanyam R, Gurikov P, Seide G, Jockenhoevel S, Ghazanfari S. Cellulose aerogel micro fibers for drug delivery applications. Mater Sci Eng C 2021; 127: 112196.
[http://dx.doi.org/10.1016/j.msec.2021.112196] [PMID: 34225849]

[42]   Ouhammou M, Nabil B, Hidar N, *et al.* Chemical activation of the skin of prickly pear fruits and cladode *Opuntia Megacantha*: Treatment of textile liquid discharges. S Afr J Chem Eng 2021; 37: 118-26.
[http://dx.doi.org/10.1016/j.sajce.2021.04.006]

[43]   Ashadujjaman M, Saifullah A, Shah DU, *et al.* Enhancing the mechanical properties of natural jute yarn suitable for structural applications. Mater Res Express 2021; 8(5): 055503.
[http://dx.doi.org/10.1088/2053-1591/abfd5e]

[44]   Zubair M, Adnan Ramzani PM, Rasool B, *et al.* Efficacy of chitosan-coated textile waste biochar applied to Cd-polluted soil for reducing Cd mobility in soil and its distribution in moringa (Moringa oleifera L.). J Environ Manage 2021; 284: 112047.
[http://dx.doi.org/10.1016/j.jenvman.2021.112047] [PMID: 33571851]

[45]   Verma M, Gahlot N, Singh SSJ, Rose NM. Enhancement of dye absorption of cotton fabric through optimization of biopolymer treatment. Cellul Chem Technol 2021; 55(3-4): 343-54.
[http://dx.doi.org/10.35812/CelluloseChemTechnol.2021.55.33]

[46]   Oikonomou EK, Messina GML, Heux L, Marletta G, Berret JF. Adsorption of a fabric conditioner on cellulose nanocrystals: synergistic effects of surfactant vesicles and polysaccharides on softness properties. Cellulose 2021; 28(4): 2551-66.
[http://dx.doi.org/10.1007/s10570-020-03672-y]

[47]   Mofradi M, Karimi H, Dashtian K, Ghaedi M. Processing Guar Gum into polyester fabric based promising mixed matrix membrane for water treatment. Carbohydr Polym 2021; 254: 116806.
[http://dx.doi.org/10.1016/j.carbpol.2020.116806] [PMID: 33357837]

[48]   Li X, Liu J, Lu Y, *et al.* Centrifugally spun starch/polyvinyl alcohol ultrafine fibrous membrane as environmentally-friendly disposable nonwoven. J Appl Polym Sci 2021; 138(40): 51169.
[http://dx.doi.org/10.1002/app.51169]

[49]   Palumbo FS, Federico S, Pitarresi G, Fiorica C, Scaffaro R, Maio A, *et al.* Effect of alkyl derivatization of gellan gum during the fabrication of electrospun membranes. J Ind Text 2021; 15280837211007508.

[50]   Mahto A, Mishra S. The removal of textile industrial Dye-RB-19 using guar gum-based adsorbent with thermodynamic and kinetic evaluation parameters. Polym Bull 2021.

[51]   Idohou EA, Fatombi JK, Osseni SA, *et al.* Preparation of activated carbon/chitosan/*Carica papaya* seeds composite for efficient adsorption of cationic dye from aqueous solution. Surf Interfaces 2020; 21: 100741.
[http://dx.doi.org/10.1016/j.surfin.2020.100741]

[52]   Naseri-Nosar M, Farzamfar S, Salehi M, Vaez A, Tajerian R, Azami M. Erythropoietin/aloe vera-releasing wet-electrospun polyvinyl alcohol/chitosan sponge-like wound dressing: *In vitro* and *in vivo* studies. J Bioact Compat Polym 2018; 33(3): 269-81.
[http://dx.doi.org/10.1177/0883911517731793]

[53]   Ramadan MA, Taha GM. EL- Mohr WZE-A. Antimicrobial and UV protection finishing of polysaccharide based textiles using biopolymer and AgNPs. Egypt J Chem 2020; 63(7): 2707-16.

[54]   Tortora M, Gherardi F, Ferrari E, Colston B. Biocleaning of starch glues from textiles by means of α-amylase-based treatments. Appl Microbiol Biotechnol 2020; 104(12): 5361-70.
[http://dx.doi.org/10.1007/s00253-020-10625-9] [PMID: 32322945]

[55]   Shahid-ul-Islam , Butola BS, Kumar A. Green chemistry based *in-situ* synthesis of silver nanoparticles for multifunctional finishing of chitosan polysaccharide modified cellulosic textile substrate. Int J Biol Macromol 2020; 152: 1135-45.
[http://dx.doi.org/10.1016/j.ijbiomac.2019.10.202] [PMID: 31783071]

[56]   Balaji AN, Ramanujam NR. Isolation and characterization of cellulose nanocrystals from Saharan aloe vera cactus fibers. Int J Polym Anal Charact 2020; 25(2): 51-64.
[http://dx.doi.org/10.1080/1023666X.2018.1478366]

[57]   Janarthanan M, Senthil Kumar M. Extraction of alginate from brown seaweeds and evolution of bioactive alginate film coated textile fabrics for wound healing application. J Ind Text 2019; 49(3): 328-51.
[http://dx.doi.org/10.1177/1528083718783331]

[58]   Jana S, Ray J, Bhanja SK, Tripathy T. Removal of textile dyes from single and ternary solutions using poly(acrylamide- *co-N* -methylacrylamide) grafted katira gum hydrogel. J Appl Polym Sci 2018; 135(10): 44849.
[http://dx.doi.org/10.1002/app.45958]

[59]   chaudhary S, Sharma J, Kaith BS, yadav S, Sharma AK, Goel A. Gum xanthan-psyllium--l-poly(acrylic acid-co-itaconic acid) based adsorbent for effective removal of cationic and anionic dyes: Adsorption isotherms, kinetics and thermodynamic studies. Ecotoxicol Environ Saf 2018; 149: 150-8.
[http://dx.doi.org/10.1016/j.ecoenv.2017.11.030] [PMID: 29156307]

[60]   Masood R, Hussain T, Miraftab M, *et al.* Novel alginate, chitosan, and psyllium composite fiber for wound-care applications. J Ind Text 2017; 47(1): 20-37.
[http://dx.doi.org/10.1177/1528083716632805]

[61]   Kolya H, Sasmal D, Tripathy T. Novel biodegradable flocculating agents based on grafted starch family for the industrial effluent treatment. J Polym Environ 2017; 25(2): 408-18.
[http://dx.doi.org/10.1007/s10924-016-0825-0]

[62]   Sahbaz DA, Acikgoz C. Adsorption of a textile dye Ostazin Black NH from aqueous solution onto

chitosan-coated perlite beads. Desalination Water Treat 2017; 67: 332-8.
[http://dx.doi.org/10.5004/dwt.2017.20376]

[63]   Darder M, Matos CRS, Aranda P, Gouveia RF, Ruiz-Hitzky E. Bionanocomposite foams based on the assembly of starch and alginate with sepiolite fibrous clay. Carbohydr Polym 2017; 157: 1933-9.
[http://dx.doi.org/10.1016/j.carbpol.2016.11.079] [PMID: 27987913]

[64]   Uyar G, Kaygusuz H, Erim FB. Methylene blue removal by alginate–clay quasi-cryogel beads. React Funct Polym 2016; 106: 1-7.
[http://dx.doi.org/10.1016/j.reactfunctpolym.2016.07.001]

CHAPTER 9

Pharmacokinetics and Toxicology of Pharmaceutical Excipients

Abstract: Progress, innovation, and development of new chemical entities fetched new defies in the drug delivery arena, and also put forward several issues including bioavailability with intestinal metabolism or efflux mechanism. However, some excipients such as surfactants have demonstrated improvement in drug bioavailability. Thus, these excipients can no longer be considered inert and require attention from a pharmaceutical regulatory perception. Biopolymers and their derivatives are gaining attention in pharmaceutical manufacturing due to their biodegradability and compatibility. However, based on the Food and Drug and Administration (FDA) guidelines, the manufacturers are required to evaluate their pharmacokinetic and toxicological properties. Several methods including Rule-of-Five and Biopharmaceutical Classification System (BCS) are used for early pharmacokinetic prediction of active and inactive pharmaceutical ingredients. Although polymers differ from therapeutic agents, similar methods can be smeared for the understanding of the absorption, distribution, metabolism, and excretion profile of bio-based pharmaceutical excipients. This chapter explores pharmacokinetic and pharmacodynamics information of biopolymers used in the design, and development of several pharmaceutical formulations.

Keywords: Biopharmaceutical classification system, Biopolymer, Lipinski rule, Pharmacokinetics.

INTRODUCTION

Nature is augmented with an extensive assortment of phytoconstituent such as biomedicine and polysaccharides that are directly or indirectly beneficial for the persistence of consistent health. Pharmaceutical excipients are traditionally known as inert components used in the formulation of pharmaceutical products that contain active constituents essential to assuring quality and safety with the effectiveness of the product [1]. Pharmaceutically inactive ingredients are the backbone of the formulation, whereas these excipients are incorporated in finished dosage forms to maintain the physicochemical stability of active pharmaceuticals during *in vitro* and *in vivo* analysis. Considering this, an excipient in pharmaceutical formulation could convert the most effective drug substance into

Sudarshan Singh & Warangkana Chunglok

an ineffectual one by modifying and conjugating the formulated product. While on the other hand, some drug molecules/products with poor pharmacokinetics can be improved by the use of excipients. Pharmaceutical excipients such as preservatives, colorants, fillers, sweeteners, binders, flavoring agents, film coating agents, encapsulating materials, lubricating agents, and their derivatives originating from plants, minerals, and animals source, are gaining importance in the development of pharmaceutical products [2]. Pharmaceutically inactive ingredients are used to expedite the pharmacodynamics and pharmacokinetics of active moiety added in the formulations. In addition, excipients also impact the physiochemical properties of medicine including consistency, solubility, and release kinetics which are a basic prerequisite for the attainment of therapeutic efficacy after administration. Natural excipients hold several advantages including biodegradability, cytocompatibility, inertness, and low cost of production over synthetic polymers. This has gained interest among pharmaceutical manufacturers for the development of various formulations as well as food products [3]. Several alternative polymers are available with pharmaceutical manufacturers that have been continuously used for the formulation of various dosage forms including solid, liquid, semisolid, *etc*. However, some of these synthetic excipients indicated side effects during *in vivo* analysis [4]. Source and the active constituent of bio-based polymers are presented in Table **1**.

Table 1. Source and active constituents of biopolymer-based polysaccharides.

Source	Composition	Properties	Reference
Buchanania lanzan Spreng	Arabinose, rhamnose, fructose, mannose, fixed oil used as permeation enhancer	Mw: 7012; ζ: 1.56 mV	[5 - 7]
Manilkara zapota (Linn.)	Rhamnose, xylose, arabinose, mannose, fructose	Mw: 2903837; ζ: 18.05 mV	[8]
Diospyros melonoxylon Roxb	Arabinose, fructose, mannose, rhamnose, xylose, and glucose	Mw: 8760; ζ: 5.90mV	[9]
Aloe vera leaves	Arabinan, arabinorhamnogalactan, galactan, galactogalacturan, glucogalactomannan, *etc*.	Mw: 30-40kDa	[10, 11]
Moringa oleifera	Glucose, rhamnose, galacturonic acid, arabinose, xylose, rhamnose	Mw: 190kDa;	[12]
Caesalpinia spinose	Galactomannans, mannose, galactose	Mw: 10^6 - 2.33 x 10^6 g/mol	[13, 14]
Prosopis juliflora	Sucrose, galactomannans, mannose, glucose, glactose	Mv x 10^6: 1.14; Viscosity: 1178 ml/g	[15 - 17]
Pithecellobium Dulce	Carbohydrate 65.79 -72.76 (%), arabinose, galactose, glucose, rhamnose	ζ: -8.6mV	[18, 19]
Annona squamosal	Carbohydrate 12.45 - 14.26 (%)	-	[20, 21]

(Table 1) cont.....

Source	Composition	Properties	Reference
Shorea wiesneri	α-resin, β-resin, dammarol acid	Average Mw: 180	[22]
Mimosa Pudica	D-xylose and D-glucuronic acid, tubulin, gallic acid, calcium oxalate crystals, Cglycosylflavones	Viscosity of 0.5% solution is 50,000 cps	[23]
Opuntia ficus indica	Arabinose, galactose, xylose, galacturonic acid, rhammnose	Mw: $15.3 - 15.7 \times 10^6$ g/mol	[24, 25]
Hibiscus rosasinensis	Rhamnose, galactose, xylose, arabinose, galacturonic acid	Viscosity: 12.234 poise	[26]
Azadirachta indica	Arabinose, galactose, xylose, rhammnose	Mw: $6.4 - 12.0 \times 10^5$; intrinsic viscosity: 2400-3600 cPs	[27]
Cassia tora Linn.	Galactomannans	Mw: 198 kDa	[28, 29]
Zizizphus jujube	Rhamnose, arabinose, glactose, glucose, xylose	-	[30, 31]
Cassia angustifolia	Rhamnose, arabinose, glactose, glucose, xylose, mannose	-	[32, 33]
Caesalpinia pulchirrima	Galactomannan	-	[34, 35]

Pharmacokinetic and Toxicology of Excipients

Pharmacokinetic and pharmacodynamics are relevant to understanding the absorption, distribution, metabolism, elimination (ADME), and therapeutic efficacy of administered medicine and/or inactive ingredient following oral, topical, intravenous, intramuscular administration. Moreover, recently a new terminology "Liberation" has been included considering studying the release kinetic process controlled by the used excipients. Toxicological screening and ADME profiling are conducted during the discovery and development of new drug entities to estimate the safety and efficacy of new drug molecules. Excipients are known as pharmacologically inactive substances and in parenthesis, their regulatory requirements to those of new drug substances require several evaluations during the development process that cost between 10 to 50 million dollars [36]. Therefore, there is a need to investigate the pharmacokinetic and pharmacodynamic profile of biopolymers as alternatives to synthetic excipients.

Basics of Pharmacokinetic and Pharmacodynamics

Pharmacokinetic and pharmacodynamics are the basis for understanding and estimating the action of active and inactive moiety administered in the body. The term pharmacokinetic is derived from the Greek word "pharmakon" and "kinetikos" meaning "drug" and "putting in motion", respectively [37]. Pharmacokinetic is also defined as a quantitative assessment of absorption, distribution, metabolism, and elimination whereas pharmacodynamics is related to the effects of the drugs and the mechanism of their action. The pharmacokinetic parameter used in the characterization of ADME is listed in Table **2**.

Table 2. Basic terminology is used in understanding the pharmacokinetics and pharmacodynamics of active and inactive moiety administered in the body.

Pharmacokinetic Terminology	Definition
Liberation	A process in which pharmaceutical products release the active medicament either in uncontrolled or in a controlled manner depending on the type of excipients used.
Absorption	Absorption of the orally administered drug is either a passive or active process in which the therapeutic agent moves from the site of administration to the site measurement.
Distribution	The movement of the product administered to and from the blood and various tissues of the body and relative proportion of the drug in the tissues.
Metabolism	A process by which the active or inactive substance administered to the body breaks down and converts it into chemical substances.
Elimination	It is a sum of processes of removing an administered drug from the body either in an un-metabolized form or by metabolic biotransformation followed by excretion.
Solubility	The amount of active or inactive pharmaceutical ingredients pass into the solution when an equilibrium is established between solute in solution.
Permeability	The capacity of active or inactive pharmaceutical ingredients to pass across the biological membrane.
Bioavailability	The rate and extent of the drug reach the systemic circulation.
Volume of distribution	Distribution of active substances within the body.
Elimination half-life	Time is required to eliminate half of the administered active substance from the body.
Clearance	The volume of plasma cleared off an active substance over a specified period.
Hepatic clearance	Clearance of a metabolized drug
Renal clearance	Elimination of un-metabolized drug

Pharmacokinetics Profiling of Active and Inactive Substances

Pharmacokinetic profiling of active or inactive pharmaceutical substances is generally processed using "rules of thumb" and a biopharmaceutical classification system (BCS) to predict the permeability and bioavailability of administered drug substances. Lipinski *et al.* [38], evaluated biologically active compounds and came up with facts related to absorption of the compound with biological properties of active or inactive substances administered in the body. Moreover,

Lipinski's rule of five is also known as the rule of thumb, which helps to determine if a biologically active chemical is likely to have the chemical and physical properties to be orally bioavailable. The thumb rule applies only to passive absorption and excludes biological transporters substrate. The basic pre-requisite of Lipinski rule of five is as:

1. No more than 5 hydrogen bond donors (articulated as the sum of all OHs and NHs)

2. No more than 10 hydrogen bond acceptors (articulated as the sum of Ns and Os)

3. Molecular mass less than 500 Da, and

4. Partition coefficient ($P_{octanol/water}$) not greater than 5

The Lipinski rule is applicable for orally administered drugs with a slight modification to predict permeation through biomembrane. Though there are numerous exceptions, Lipinski's Rule-of-Five still provides valuable information on the physicochemical property of either active or inactive substance interaction with the body (Table 3).

Table 3. Desecration of 2 or more of rule-of-thumb conditions predictions of orally and non-orally administered molecule [39].

Route Of Administered	Mw (Da)	H-Bound Donors	H Accepters	Log P
Ophthalmic	500	3	8	4.2
Inhalation	500	4	10	3.4
Transdermal	335	2	5	5.0
Orally	500	5	10	5.0

The biopharmaceutical classification system concept is continuously used as a tool in pharmaceutical science and technology to classify compounds based on their solubility and permeability (Fig. 1). In 1995, Amidon and coworkers introduced the biopharmaceutical classification system as a useful and alternative tool for the assessment of bioequivalence study utilizing *in vitro* dissolution test compared with innovator formulation [40]. The manufacturers in the application of new drug approvals, abbreviated new drug applications and are now continuously implementing the principles of BCS and post-approval changes in pharmaceuticals. Moreover, BCS has been mandated by food drug administration as a regulatory guideline considering waiver of *in vivo* bioavailability and bioequivalence studies for immediate-release oral solid dosage [41].

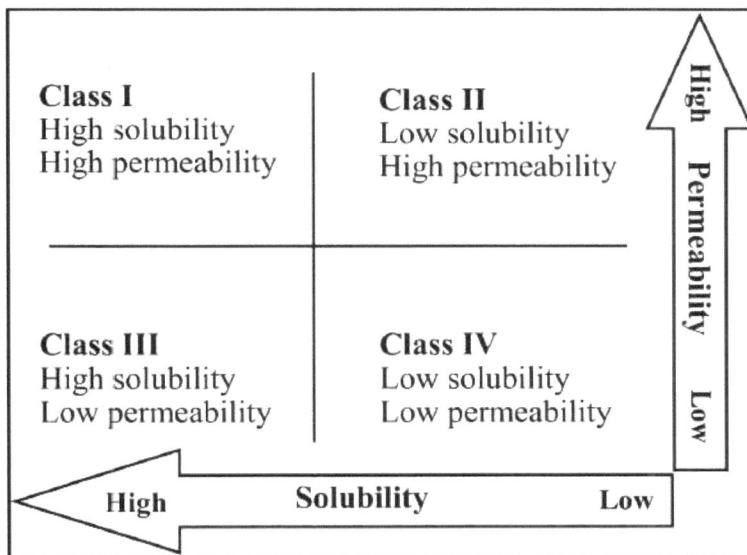

Fig. (1). The biopharmaceutical classification system in formulation development.

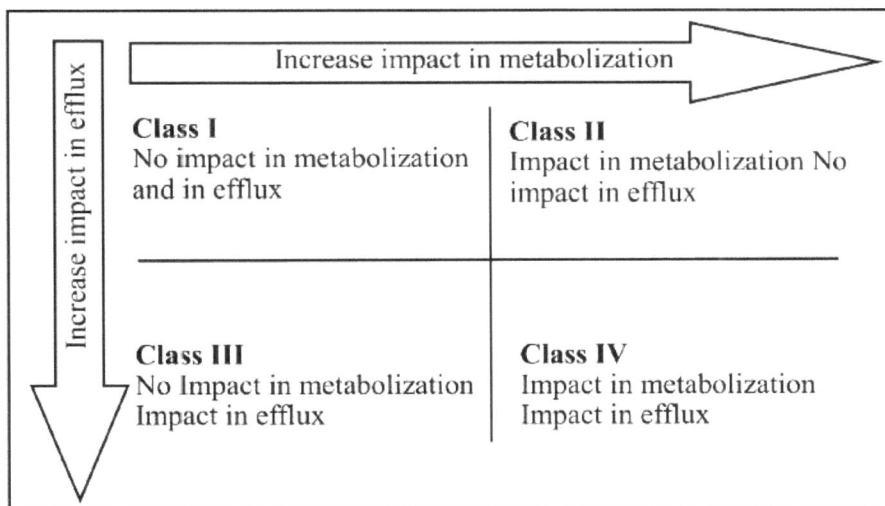

Fig. (2). The biopharmaceutical classification system of excipients [1].

The menace linked with the active pharmaceutical substances relating to safety, quality, and efficacy issues has been securely controlled and well established. Similarly, for an excipient, safety guidelines are required to ensure formulation quality and efficacy with no effect on the drug incorporated. Vasconcelos and co-workers reported for the first time, four classes of a biopharmaceutical classification system for excipients (BCSE). According to BCSE, the excipients were classified into four classes considering metabolization and efflux mechanism (Fig. **2**). Class, I represents a low risk of impact on drug safety and efficacy and includes excipients with a low risk of metabolization and efflux mechanism interaction. Class II and III include excipients of high risk, that interfere with intestinal metabolization, however this class of excipients does not interfere with the efflux mechanism. The third class of BCSE includes excipients with no impact on metabolism but influences efflux mechanism. Class IV excipients are considered the riskiest, impacting both metabolization and an efflux mechanism. The relationship between key molecular properties of various classes of excipients that affect the ADME of administered substances is presented in Table **4**. In addition, physicochemical and pharmacokinetic properties of some bio-based polymeric inactive ingredients are presented in Table **5**. Furthermore, a summary of the classification of the current state of art and knowledge about the excipients is presented in Fig. (**3**).

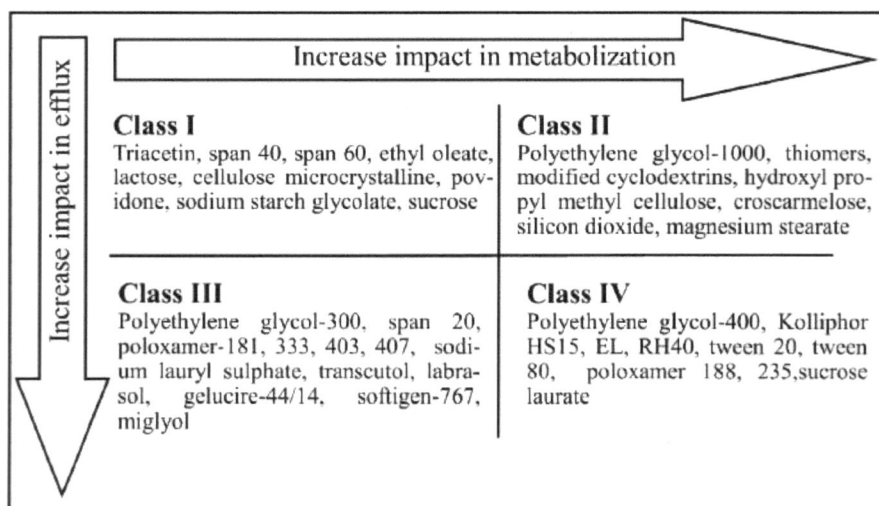

Increase impact in efflux

Increase impact in metabolization

Class I
Triacetin, span 40, span 60, ethyl oleate, lactose, cellulose microcrystalline, povidone, sodium starch glycolate, sucrose

Class II
Polyethylene glycol-1000, thiomers, modified cyclodextrins, hydroxyl propyl methyl cellulose, croscarmelose, silicon dioxide, magnesium stearate

Class III
Polyethylene glycol-300, span 20, poloxamer-181, 333, 403, 407, sodium lauryl sulphate, transcutol, labrasol, gelucire-44/14, softigen-767, miglyol

Class IV
Polyethylene glycol-400, Kolliphor HS15, EL, RH40, tween 20, tween 80, poloxamer 188, 235,sucrose laurate

Fig. (3). Recently included synthetic excipients in the biopharmaceutical classification system of excipients [1].

Table 4. Relationship between key molecular properties that affect adme of administered substances [42].

Parameters	Neutral Molecules		Basic Molecules		Acidic Molecules		Zwitterion Molecules	
	Mw < 400, Clogp < 4	Mw > 400, Clogp > 4	Mw < 400, Clogp < 4	Mw > 400, Clogp > 4	Mw < 400, Clogp < 4	Mw > 400, Clogp > 4	Mw < 400, Clogp < 4	Mw > 400, Clogp > 4
Solubility	Average	Lower	Higher/average	Lower/average	Higher	Average/higher	Higher	Average/higher
Permeability	Higher	Average/higher	Higher/average	Average	Lower	Average/lower	Lower	Lower/average
Bioavailability	Average	Lower	Average	Lower	Average	Average	Lower	Lower

(Table 4) cont.....

Parameters	Neutral Molecules		Basic Molecules		Acidic Molecules		Zwitterion Molecules	
	Mw < 400, Clogp < 4	Mw > 400, Clogp > 4	Mw < 400, Clogp < 4	Mw > 400, Clogp > 4	Mw < 400, Clogp < 4	Mw > 400, Clogp > 4	Mw < 400, Clogp < 4	Mw > 400, Clogp > 4
Volume of distribution	Average	Average	Average	Average	Lower	Lower	Lower	Average/lower
Plasma protein binding	Average	Higher	Lower	Average	Average/higher	Higher	Average/lower	Higher
CNS penetration	Higher/average	Average/lower	Higher/average	Average/lower	Lower	Lower	Average/lower	Lower
Brain tissue binding	Lower	Higher	Lower	Higher	Lower	Higher	Lower	Higher
P-glycoprotein efflux	Average	Higher/average	Average	Higher/average	Lower	Lower	Average	Average
In vivo clearance	Average	Average	Average	Higher/average	Lower/average	Average	Average	Average
hERG inhibition	Lower	Lower	Average/higher	Higher	Lower	Lower	Lower	Average/lower
P450 inhibition	Lower 2C9, 2C19, 2D6 and 3A4 inhibition	Higher 2C9, 2C19, and 3A4 inhibition	Lower 1A2, 2C9, and C19 inhibition	Lower 1A2 inhibition	Lower 1A2, 2C9, 2C19, 2D6, and 3A4 inhalation	Lower 1A2, 2C19, 2D6, & 3A4 inhibition	Lower 1A2, 2C9, 2C19, 2D6, and 3A4 inhibition	Lower 1A2, 2C19, and 3A4 inhibition
	Higher 1A2 inhibition	Lower 1A2 inhibition	Average 2D6 and 3A4 inhibition	Average 2C9, 2C19, inhibition		Higher 2C9 inhibition		Average 2C9, 2D6 inhibition
		Average 2D6 inhibition		Higher 2D6 and 3A4 inhibition				

Table 5. Physicochemical and pharmacokinetic properties of some bio-based polymeric inactive ingredients.

Polysaccharides Active Content	Solubility (Mg/Ml)	Log *P*	Bioavailability (%)	Reference
Aspartame	10	0.5	Well absorb from Gut, undergoes hepatic metabolism	[43]
β-Carotene	< 0.001	15	22	[44]
Curcumin	0.0003	3.1	< 10	[45 - 47]
Glucose	640	-3.3	100	[48 - 50]
Lycopene	< 0.001	15	< 2	[44]
Maltose	520	-4.8	0	[50, 51]
Mannitol	650	-3.3	7-40	[50, 52]
Saccharin	3	0.9	≥ 85	[53, 54]
Sorbitol	770	-3.3	25-90	[50, 55]
Sucralose	300	0.2	< 15	[56]
Menthol	0.5	3.2	< 90	[57]
Glucuronic acid	8.58	2.34	0.56	[58]
Arabinose	8.58	2.32	0.55	[58]

(Table 5) cont.....

Polysaccharides Active Content	Solubility (Mg/Ml)	Log *P*	Bioavailability (%)	Reference
Rhamnose	4.72	2.09	0.55	[58]
Xylitol	1.68	2.48	0.55	[58]
Carvone	5.81	2.71	0.55	[58]
Limonene	4.33	4.57	0.55	[58]
Myrcene	1.22	4.17	-0.55	[58]
Thymol	9.74	3.30	0.55	[58]
Cymene	2.08	3.30	0.55	[58]

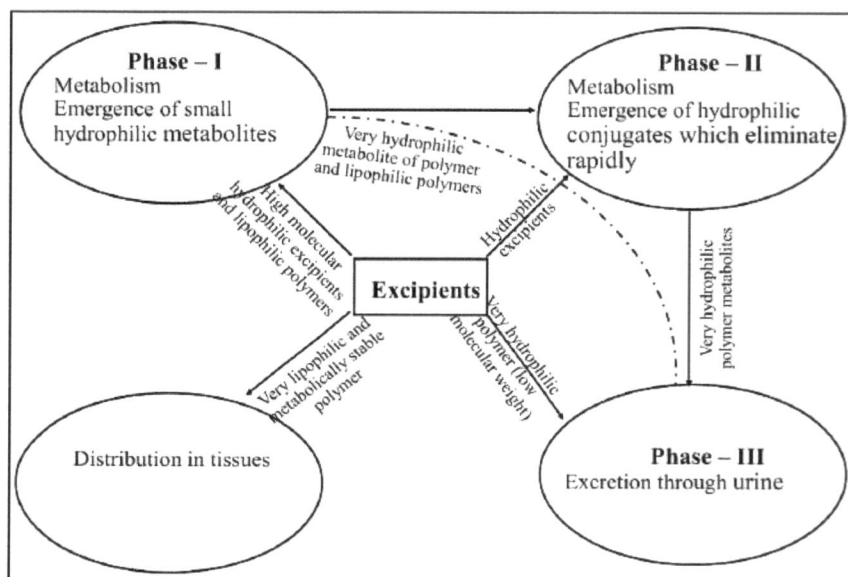

Fig. (4). Absorption, distribution, metabolic, and excretion pathway of administered pharmaceutical excipients, whereas phase I indicates the increment in hydrophilicity of the excipient to facilitate phase II conjugations, and phase III represents the elimination of either unmetabolized or metabolized inactive substances [61].

The pharmacokinetics of active pharmaceutical ingredients can be improved *via* physicochemical modification. Although modifications can improve the ADME of the drug, such active ingredients require being significantly lipophilic to permeate biological membranes and provoke a therapeutic response. Bodor and coworkers introduced the concept of soft drug design in 2004 which improved the pharmacokinetics and pharmacodynamics of the controlled metabolic pathway [59]. Similar to soft drug innovation, the design of soft excipients such as antimicrobial was introduced by Thorsteinsson and coworkers [60]. Generally, excipients are considered biologically inactive, inert, unable to interact with drug receptors to produce a pharmacological response, and unable to amend enzymatic action. However, considering soft excipients modification theory, there are some exceptions available that need to be assessed by the researcher [61]. Moreover, physicochemical amendment of excipients alters the lipophilicity and molecular weight, causing significant alteration in solubility, permeability, and excretion rate as represented in Fig. (**4**).

CONCLUSIONS AND FUTURE PERSPECTIVES

Pharmacokinetics and pharmacodynamics of active and inactive materials administered are an important pharmacological constraint that depends on solubility and permeability. This chapter elaborated on the concept of Lipinski's rule of five and BCS, which established a new concept stating that excipients are not inactive substances. In the future, excipients might be considered an integral part of the dosage form that could significantly ameliorate various pharmacokinetic and pharmacodynamics phenomena of drug substances.

REFERENCE

[1]   Vasconcelos T, Marques S, Sarmento B. The biopharmaceutical classification system of excipients. Ther Deliv 2017; 8(2): 65-78.
        [http://dx.doi.org/10.4155/tde-2016-0067] [PMID: 28088879]

[2]   Sabalingam S, Jayasuriya WJAB. Pharmaceutical excipients of marine and animal origin: a review. Biol Chem Res 2019; 6: 184-96.

[3]   Heer D. Novel excipients as different polymers: a review. J Drug Deliv Therap 2013; p. 3.

[4]   Singh PMG, Dinda SC. Natural excipients in pharmaceutical formulations. In: Mandal SC, Chakraborty R, Sen S, Eds. Evidence based validation of traditional medicines. Singapore: Springer 2021; pp. 829-69.
        [http://dx.doi.org/10.1007/978-981-15-8127-4_40]

[5]   Singh S, Bothara S. Morphological, physico-chemical and structural characterization of mucilage isolated from the seeds of *Buchanania lanzan* Spreng. Int J Health Allied Sci 2014; 3(1): 33.
        [http://dx.doi.org/10.4103/2278-344X.130609]

[6]   Singh S, Sunil BB, Patel DR, Mahobia KN. Formulation and *ex-vivo* evaluation of transdermal patches of glipizide using the penetration enhancer *Buchanania lanzan* (Spreng) seed oil. Int J Pharm Sci Nanotech 2015; 8(1)

[7]   Bothara Sunil B, Singh S. Fatty Acid Profile of *Buchanania lanzan* Spreng seed oil by gas chromatography-mass spectrometry. Inventi Rapid: Pharm Ana Qual Assur 2011.

[8]   Singh S, Bothara SB. *Manilkara zapota* (Linn.) Seeds: A potential source of natural gum. ISRN Pharm 2014; 2014: 1-10.
        [http://dx.doi.org/10.1155/2014/647174] [PMID: 24729907]

[9]   Singh S, Bothara SB. Physico-chemical and structural characterization of mucilage isolated from seeds of *Diospyros melonoxylon* Roxb. Braz J Pharm Sci 2014; 50(4): 713-25.
        [http://dx.doi.org/10.1590/S1984-82502014000400006]

[10]   Maru SGSS. Physicochemical and mucoadhesive strength characterization of natural polymer obtained from leaves of *Aloe vera.* Pharmtechmedica 2013; 2(3): 303-8.

[11]   Ni Y, Turner D, Yates Ka, Tizard I. Isolation and characterization of structural components of *Aloe*

vera L. leaf pulp. Int immunopharm 2004; 4(14): 1745-55.

[12]   Raja W, Bera K, Ray B. Polysaccharides from *Moringa oleifera* gum: structural elements, interaction with β-lactoglobulin and antioxidative activity. RSC Advances 2016; 6(79): 75699-706.
[http://dx.doi.org/10.1039/C6RA13279K]

[13]   Picout DR, Ross-Murphy SB, Jumel K, Harding SE. Pressure cell assisted solution characterization of polysaccharides. 2. Locust bean gum and tara gum. Biomacromolecules 2002; 3(4): 761-7.
[http://dx.doi.org/10.1021/bm025517c] [PMID: 12099820]

[14]   Daas PJH, Schols HA, de Jongh HHJ. On the galactosyl distribution of commercial galactomannans. Carbohydr Res 2000; 329(3): 609-19.
[http://dx.doi.org/10.1016/S0008-6215(00)00209-3] [PMID: 11128589]

[15]   Singh S, Sheth NR, Rajesh D. Pharmaceutical characterization of *Prosopis juliflora* (Sw) seed mucilage-excipient. Acta Pharma Sciencia 2010; 52(4): 487-94.

[16]   Marangoni A, Alli I. Composition and properties of seeds and pods of the tree legume *Prosopis juliflora* (DC). J Sci Food Agric 2006; 44: 99-110.

[17]   Rincón F, Muñoz J, Ramírez P, Galán H, Alfaro MC. Physicochemical and rheological characterization of *Prosopis juliflora* seed gum aqueous dispersions. Food Hydrocoll 2014; 35: 348-57.
[http://dx.doi.org/10.1016/j.foodhyd.2013.06.013]

[18]   Chaudhari BB, Annapure US. Physiochemical and rheological characterization of *pithecellobium dulce* (Roxb.) benth gum exudate as a potential wall material for the encapsulation of rosemary oil. Carbohydrate Polymer Technologies and Applications 2020; 1: 100005.
[http://dx.doi.org/10.1016/j.carpta.2020.100005]

[19]   Singh S, Goswami H. Formulation of oral mucoadhesive tablets of pioglitazone using natural gum from seeds of *Pithecellobium dulce*. International Journal of Pharmaceutical Sciences and Nanotechnology 2015; 8(4): 3031-8.
[http://dx.doi.org/10.37285/ijpsn.2015.8.4.6]

[20]   Hassan LG, Muhammad M, Umar KJ, Sokoto A. Comparative study on the proximate and mineral contents of the seed and pulp of sugar Apple (*Annona squamosa*). Nigerian J Basic Applied Sci 2008; 16: 179-82.

[21]   Singh S, Santoki R. Development and appraisal of mucoadhesive tablets of hydralazine using isolated mucilage of *Annona squamosa* seeds. Int J Pharm Sci Nanotech 2016; 9(4): 3349-56.

[22]   Morkhade DM, Fulzele SV, Satturwar PM, Joshi SB. Gum copal and gum damar: Novel matrix forming materials for sustained drug delivery. Indian J Pharm Sci 2006; 68(1): 53.
[http://dx.doi.org/10.4103/0250-474X.22964]

[23]   Singh K, Kumar A, Langyan N, Ahuja M. Evaluation of *Mimosa pudica* seed mucilage as sustained-release excipient. AAPS PharmSciTech 2009; 10(4): 1121-7.
[http://dx.doi.org/10.1208/s12249-009-9307-1] [PMID: 19763837]

[24]   Matsuhiro B, Lillo L, Saenz C, Urzúa C, Zárate O. Chemical characterization of the mucilage from fruits of Opuntia ficus indica. Carbohyd Polym 2006; p. 63.

[25]   Felkai-Haddache L, Remini H, Dulong V, *et al.* conventional and microwave-assisted extraction of mucilage from *Opuntia ficus*-indica Cladodes: physico-chemical and rheological properties. Food Bioprocess Technol 2016; 9(3): 481-92.
[http://dx.doi.org/10.1007/s11947-015-1640-7]

[26]   Kassakul W, Praznik W, Viernstein H, Hongwiset D, Phrutivorapongkul A, Leelapornpisid P, *et al.* Characterization of the mucilage's extracted from Hibiscus rosasinensis linn and Hibiscus mutabilis linn and their skin moisturizing effect. Int J Pharmcy Pharm Sci 2014; p. 6.

[27]   Mawahib E. Moniem EAH, Mohammed E Osman. Characterization and rheological behavior of neem

gum (*Azadirachta indica*). Int J Chem Stud 2018; 6(3): 1977-81.

[28]   Singh S, Bothara S, Singh S, Patel RD, Mahobia NK. Pharmaceutical characterization of *Cassia tora* of seed mucilage in tablet formulations. Pharm Lett 2010; 2(5): 54-61.

[29]   Pawar HA, Lalitha KG. Extraction, characterization, and molecular weight determination of *Senna tora* (L.) Seed Polysaccharide. Int J Biomater 2015; 2015: 1-6.
        [http://dx.doi.org/10.1155/2015/928679] [PMID: 26640490]

[30]   Zhao Z, Liu M, Tu P. Characterization of water soluble polysaccharides from organs of Chinese Jujube (Ziziphus jujuba Mill. cv. Dongzao). Eur Food Res Technol 2008; 226(5): 985-9.
        [http://dx.doi.org/10.1007/s00217-007-0620-1]

[31]   Singh SK, Gendle R, Sheth NR, Roshan P, Singh S. Isolation and evaluation of binding property of *Ziziphus jujube* lamx. seed mucilage in tablet formulations. J Glob Pharma Technol 2010; 2(1): 98-102.

[32]   Müller B, Kraus J, Franz G. Chemical structure and biological activity of water-soluble polysaccharides from *Cassia angustifolia* leaves. Planta Med 1989; 55(6): 536-9.
        [http://dx.doi.org/10.1055/s-2006-962088] [PMID: 2616672]

[33]   Singh S, Singh S. Isolation and evaluation of *Cassia aungostifolia* seed mucilage as granulating agent. Int J Pharm Sci Res 2010; 1(8): 118-25.

[34]   Stella Regina AM. VAdO, Amanda Mazza C de O, Marjory Lima HA, Judith Pessoa de Andrade F, Regina Celia Monteiro de P, Felipe Domingos de S, Ana Cristina de Oliveira MM, Frederico José B, Renato de Azevedo M. *Caesalpinia pulcherrima* seed galactomannan on rheological properties of dairy desserts. Cienc Rural 2020; 50(6): e20190176.

[35]   Singh S, Sangeeta S, Bothara SB, Roshan P. Pharmaceutical characterization of some natural excipient as potential mucoadhesives agent. Pharma Res J 2010; 4: 91-104.

[36]   IEPC. IEPC Americas and IQ Consortium push for novel excipient review pathway. IPEC e-newsletter - excipients Insight 2014. Available from: http://ipec-europe.org/newsletter.asp%3Fnlaid=638& nlid=647

[37]   Smith Y. Pharmacokinetics. News medical life science 2015; 1-4. Available from: https://www.news-medical.net/health/Pharmacokinetics.aspx

[38]   Lipinski CA, Lombardo F, Dominy BW, Feeney PJ. Experimental and computational approaches to estimate solubility and permeability in drug discovery and development settings 1PII of original article: S0169-409X(96)00423-1. The article was originally published in Advanced Drug Delivery Reviews 23 (1997) 3–25. 1. Adv Drug Deliv Rev 2001; 46(1-3): 3-26.
        [http://dx.doi.org/10.1016/S0169-409X(00)00129-0] [PMID: 11259830]

[39]   Choy YB, Prausnitz MR. The rule of five for non-oral routes of drug delivery: ophthalmic, inhalation and transdermal. Pharm Res 2011; 28(5): 943-8.
        [http://dx.doi.org/10.1007/s11095-010-0292-6] [PMID: 20967491]

[40]   Amidon GL, Lennernäs H, Shah VP, Crison JR. A theoretical basis for a biopharmaceutic drug classification: the correlation of *in vitro* drug product dissolution and *in vivo* bioavailability. Pharm Res 1995; 12(3): 413-20.
        [http://dx.doi.org/10.1023/A:1016212804288] [PMID: 7617530]

[41]   Sullivan JO, Blake K, Berntgen M, Salmonson T, Welink J, Pharmacokinetics Working P. Overview of the European medicines agency's development of product-specific bioequivalence guidelines. Clin Pharmacol Ther 2018; 104(3): 539-45.
        [http://dx.doi.org/10.1002/cpt.957] [PMID: 29319156]

[42]   Gleeson MP. Generation of a set of simple, interpretable ADMET rules of thumb. J Med Chem 2008; 51(4): 817-34.
        [http://dx.doi.org/10.1021/jm701122q] [PMID: 18232648]

[43]   Puthrasingam S, Heybroek WM, Johnston A, *et al.* Aspartame pharmacokinetics - the effect of ageing. Age Ageing 1996; 25(3): 217-20.
[http://dx.doi.org/10.1093/ageing/25.3.217] [PMID: 8670556]

[44]   Burri BJ, Neidlinger TR, Clifford AJ. Serum carotenoid depletion follows first-order kinetics in healthy adult women fed naturally low carotenoid diets. J Nutr 2001; 131(8): 2096-100.
[http://dx.doi.org/10.1093/jn/131.8.2096] [PMID: 11481400]

[45]   Heger M, van Golen RF, Broekgaarden M, Michel MC. The molecular basis for the pharmacokinetics and pharmacodynamics of curcumin and its metabolites in relation to cancer. Pharmacol Rev 2014; 66(1): 222-307.
[http://dx.doi.org/10.1124/pr.110.004044] [PMID: 24368738]

[46]   Prasad S, Tyagi AK, Aggarwal BB. Recent developments in delivery, bioavailability, absorption and metabolism of curcumin: the golden pigment from golden spice. Cancer Res Treat 2014; 46(1): 2-18.
[http://dx.doi.org/10.4143/crt.2014.46.1.2] [PMID: 24520218]

[47]   Pandey MK, Kumar S, Thimmulappa RK, Parmar VS, Biswal S, Watterson AC. Design, synthesis and evaluation of novel PEGylated curcumin analogs as potent Nrf2 activators in human bronchial epithelial cells. Eur J Pharm Sci 2011; 43(1-2): 16-24.
[http://dx.doi.org/10.1016/j.ejps.2011.03.003] [PMID: 21426935]

[48]   Hahn RG, Ljunggren S, Larsen F, Nyström T. A simple intravenous glucose tolerance test for assessment of insulin sensitivity. Theor Biol Med Model 2011; 8(1): 12.
[http://dx.doi.org/10.1186/1742-4682-8-12] [PMID: 21535887]

[49]   Heller N, Kalant N, Hoffman MM. The relationship between insulin responsiveness and blood glucose half-life in normal and diabetic subjects. J Lab Clin Med 1958; 52(3): 394-401.
[PMID: 13575929]

[50]   Bouchard A, Hofland GW, Witkamp GJ. Properties of sugar, polyol, and polysaccharide water–ethanol solutions. J Chem Eng Data 2007; 52(5): 1838-42.
[http://dx.doi.org/10.1021/je700190m]

[51]   Tahara Y, Fukuda M, Yamamoto Y, *et al.* Metabolism of intravenously administered maltose in renal tubules in humans. Am J Clin Nutr 1990; 52(4): 689-93.
[http://dx.doi.org/10.1093/ajcn/52.4.689] [PMID: 2403061]

[52]   Cloyd JC, Snyder BD, Cleeremans B, Bundlie SR, Blomquist CH, Lakatua DJ. Mannitol pharmacokinetics and serum osmolality in dogs and humans. J Pharmacol Exp Ther 1986; 236(2): 301-6.
[PMID: 3080582]

[53]   Colburn WA, Bekersky I, Blumenthal HP. A preliminary report on the pharmacokinetics of saccharin in man: single oral dose administration. J Clin Pharmacol 1981; 21(4): 147-51.
[http://dx.doi.org/10.1002/j.1552-4604.1981.tb05692.x] [PMID: 7240436]

[54]   Sweatman TW, Renwick AG, Burgess CD. The pharmacokinetics of saccharin in man. Xenobiotica 1981; 11(8): 531-40.
[http://dx.doi.org/10.3109/00498258109045864] [PMID: 7303723]

[55]   Nau R, Dreyhaupt T, Kolenda H, Prange HW. Low blood-to-cerebrospinal fluid passage of sorbitol after intravenous infusion. Stroke 1992; 23(9): 1276-9.
[http://dx.doi.org/10.1161/01.STR.23.9.1276] [PMID: 1519282]

[56]   Roberts A, Renwick AG, Sims J, Snodin DJ. Sucralose metabolism and pharmacokinetics in man. Food Chem Toxicol 2000; 38 (Suppl. 2): 31-41.
[http://dx.doi.org/10.1016/S0278-6915(00)00026-0] [PMID: 10882816]

[57]   Martin D, Valdez J, Boren J, Mayersohn M. Dermal absorption of camphor, menthol, and methyl salicylate in humans. J Clin Pharmacol 2004; 44(10): 1151-7.
[http://dx.doi.org/10.1177/0091270004268409] [PMID: 15342616]

[58]   Arora K, Singh S. *In-silico* bioavailability and pharmacokinetic study of some herbal mucilages. Int J Res Pharm Nano Scie 2020; 9(6): 299-309.

[59]   Bodor N, Buchwald P. Designing safer (soft) drugs by avoiding the formation of toxic and oxidative metabolites. Mol Biotechnol 2004; 26(2): 123-32.
[http://dx.doi.org/10.1385/MB:26:2:123] [PMID: 14764938]

[60]   Thorsteinsson T, Másson M, Kristinsson KG, Hjálmarsdóttir MA, Hilmarsson H, Loftsson T. Soft antimicrobial agents: synthesis and activity of labile environmentally friendly long chain quaternary ammonium compounds. J Med Chem 2003; 46(19): 4173-81.
[http://dx.doi.org/10.1021/jm030829z] [PMID: 12954069]

[61]   Loftsson T. Excipient pharmacokinetics and profiling. Int J Pharm 2015; 480(1-2): 48-54.
[http://dx.doi.org/10.1016/j.ijpharm.2015.01.022] [PMID: 25596414]

CHAPTER 10

Bibliometric Analysis of Bio-Based Pharmaceutical Excipients

Abstract: Recently bibliometric analysis has gained significant importance in quantitative assessment for analyzing scientific outputs, the linkage between universities, authors, funding organizations, and development enactment, with several other applications. Therefore, the scientific community needs an advanced tool to analyze a wide range of scientific data with precision and accuracy. This chapter aims to provide up-to-date bibliometric analysis on bio-based pharmaceutical excipients including network and overlay visualization for publication from 2000 to 2021, retrieved from the Scopus database. The documents considered were original research and conference proceedings numbering 2923. The bibliometric analysis revealed that research interests in bio-based are expanding throughout the globe, as a potential source of biomaterial for allied pharmaceutical sciences.

Keywords: Bibliometric analysis, Bio-based polymer, Bio-based polysaccharides, Natural polysaccharides, Plant mucilage.

INTRODUCTION

A comprehensive understanding of the physicochemical properties of inactive pharmaceuticals is necessary to develop quality pharmaceutical products and promote the required therapeutic efficacy. In both conventional and modified pharmaceuticals, active pharmaceuticals have been reported to be incompatible in various instances with several synthetic excipients including bulk pharmaceuticals [1, 2]. Synthetic excipients have been extensively modified to minimize issues related to incompatibility, however, due to several drawbacks including biodegradability, compatibility, and physicochemical properties of active pharmaceuticals, poor bioavailability is observed. Thus, the development of natural pharmaceutical excipients with multifunctional properties is an urgent need.

Polymers are pharmaceuticals that serve as a backbone for the development of several drug delivery categories including targeted delivery, tissue engineering, modified drug delivery, probiotics, *etc.* Both synthetic and semisynthetic excipients play important roles in the formulation and development of numerous

Sudarshan Singh & Warangkana Chunglok

dosages forms labelled as finished pharmaceuticals. The increase in dependence towards synthetic and semisynthetic polymers is partly due to the datum of unconstrained and cost-effective possibilities available for amendment in the basic skeleton to develop products with exact requisites. Subsequently, many of the widely used pharmaceutical excipients are obtained by amendments in some natural derivatives to improve their functionality. However, despite several benefits with synthetic and semisynthetic excipients, pharmaceutical manufacturers need to address various challenges before their expansion and utility can be accepted effectively. Biological macromolecules have gained significant importance due to their biodegradability with ease of availability and multi-functionality. Bio-based polysaccharides can be obtained from plants or other living things, and agriculture resources. Natural polysaccharides are macromolecules composed of repeating monomeric units interlinked by covalent bonding. The word "polymer" derived from the Greek term *polus,* which means much or many whereas *meros*, means part. The word was first devised by Jons Jacob Berzelius with a distinct definition from the IUPAC nomenclature definition [3].

Bio-based pharmaceutical excipients are one of the fastest-growing products within the pharmaceutical market due to their sustainability including biodegradability, biocompatibility, low production cost, with abundance in availability, compared to synthetic and semisynthetic polymers. Moreover, biological macromolecules originating from natural resources are renewable and proper harvesting with cultivation can provide a constant supply. The current chapter focuses on the bibliometric analysis of bio-based excipients with pharmaceutical applications (Table 1). Several reviews and monographs with scientific data are available on natural excipients, however a comprehensive bibliometric analysis of natural originating pharmaceutical inactive ingredients is lacking. The content presented in this chapter provides the reader a brief on scientific data available from the last two decades with a considerable preference on the keywords, major funding associated with bio-based polymer research, the linkage between potential authors, and leading countries with maximum citations on bio-based polymer research.

Table 1. Bio-based excipients with pharmaceutical applications.

Polysaccharides	Application	Reference
Photo-cross linkable alginic acid and hyaluronic acid	*In situ* photo-polymerization	[4]
Alginate from brown algae	Pharmaceutical microsphere	[5, 6]
Guar gum	Deflocculants	[7]
Graft vinyl monomers with chitosan	Pharmaceutical excipients	[8]

Polysaccharides	Application	Reference
Fenugreek mucilage	Flocculating agent	[9]
Plantago psyllium mucilage	Pharmaceutical excipients	[10, 11]
Chitosan and alginic acid	Gel beads	[12]
Hyaluronan	Biomedical application	[13]
Lacquer polysaccharide	Anti-tumor efficacy	[14]
Misgurnus anguillicudatus polysaccharide	Immunomodulation	[15]
Cassava starch polysaccharide	Packaging	[16]
Apple pectin	Sorption properties	[17, 18]
Tamarindus indica pods seeds mucilage	Food grade flocculant	[19 - 24]
Coccinia indica fruits mucilage	Removal of dyes	[25]
Conyza canadensis polysaccharides	Anti-aggregatory and anti-oxidant	[26]
Psyllium husk	Colon specific drug delivery	[27 - 29]
Pullulan	Nanoparticles	[30]
Hibiscus esculentus and *Trigonella foenum graceum*	Pharmaceutical flocculants	[31]
Gellan gum	Injectable vehicle and bead	[32, 33]
Cassia auriculata seed mucilage	Tablet binder	[34]
Zizyphus jujuba lamk. Seed mucilage	Tablet binder	[35]
Cassia tora of seed mucilage	Tablet binder	[36]
Cassia sophera Linn seed mucilage	Tablet binder	[37]
Prosopis juliflora seed mucilage	Tablet binder	[38]
Cassia aungostifolia seed mucilage	Tablet binder	[39]
Caesalpinia pulchirrima and *Leucaena leucocephala* seed mucilage	Mucoadhesion	[40]
Cassia fistula seed mucilage	Tablet binder	[41]
Diospyros melonoxylon Roxb, *Buchanania lanzan* Spreng, and *Manilkara zapota* (Linn.) P. Royen syn seed mucilage	Mucoadhesion	[42 - 51]
Aloe vera (L.) Burm. f leaves mucilage	Mucoadhesion	[52, 53]
Pithecellobium dulce seed mucilage	Mucoadhesion	[54]
Annona squamosa (L.) Burm. fruit mucilage	Mucoadhesion	[55]
Konjac-graft-poly (acrylamide)-co-sodium xanthate	Flocculant	[56]
Schizophyllan polysaccharides from fungi Schizophyllum	Collagen	[57]
Lycium barbarum L polysaccharides	Antioxidant	[58]
Bora rice starch	Pharmaceutical excipients	[59, 60]
Xanthan gum, guar gum, and chitosan	Colon drug delivery	[61]
Okra gum	Sustained-release tablets	[62]

(Table 1) cont.....

Polysaccharides	Application	Reference
Locust bean gum, guar gum, xanthan gum, and tamarind gum	Adhesives	[63]
Isabgol husk	Hydrogel	[64]
Alginic acid from *Laminaria digitate*, and *Ascophyllum nodosum*	Pharmaceutical excipients	[65]
Pectin	Targeted drug delivery	[66 - 68]
Cassia grandis seeds galactomannan	Hydrogel	[69]
Hyaluronan	Hydrogel	[70]
Arabinogalactan from portulaca	Synthesis of AgNPs	[71]
Fish gelatin	Nanoparticles	[72]
Sodium alginate and gum Arabic	Synthesis of AgNPs	[73]
Fenugreek gum and xyloglucan	Pharmaceutical excipients	[74]
Pectin grafted with polyhydroxybutyrate	Scaffold materials	[75]
Konjac glucomannan	Pharmaceutical excipients	[76 - 78]
karaya gum grafted with acrylic acid	Hydrogel	[79]
K-carrageenan and pectin	Food chemistry	[80]
Tragacanth Gum	Nano fiber	[81]
Graft Aloe vera polysaccharide	Pharmaceutical excipients	[82]
Bletilla striata polysaccharide	Pharmaceutical excipients	[83]
Commiphora myrrha polysaccharide	Wound dressings	[84]
Punica granatum fruit polysaccharide	Fluorescence imaging	[85]
Schizophyllan commune polysaccharide	Enhanced oil recovery	[86]

Study Design

This study was reported by Preferred Reporting Items for Systematic Reviews and Meta-Analyses (PRISMA) extension statement for network meta-analysis [87].

Data Collection

The data set presented in this study was obtained from the online 'Scopus' database which has approximately 69 million records with access to *Elsevier* and authors citation database and Web of Science [88]. In addition, to maintain consistency and to avoid data overlap, data sets were collected simultaneously. Initially, data was obtained from the Web of Science and Scopus to compare the volume of the material available on each database. The Scopus database had 2923 published scientific papers related to the topic, while the Web of Science yielded 994 published research papers. The data retrieval steps and the inclusion and exclusion criteria are depicted in Fig. (1). In this study, the data set was collected from 2000 to 2020 and accessed on 16 February 2021. Moreover, a longer time duration was required to understand the research progression during the last few decades on bio-based pharmaceutical excipients, with the hope of yielding thought-provoking results. Data were retrieved using "bio-based polymers", "bio based polysaccharides", and "natural polysaccharides" as core keywords. The strategic search for retrieving original research published during 2000 to 2020 is as follows: ("bio-based polymer*" or "bio-degradable polymer*" or "natural polysaccharides*") AND ("bio based polymer" or "bio degradable polymer" or "biobased polymer" or "biodegradable polymer") were entered with the specific document type. The quotation marks (" ") were used for specifying the exact required phrase, whereas the asterisk (*) was a shortcut for retrieving both the singular and plural versions of required keywords. Document type was limited to Original Article or Conference Proceedings Paper and Publication year ranged from 2000 – 2020, to internment research from beginning. The 2923 data obtained from the Scopus database were imported into VOSviewer (version 1.6.16) software and Microsoft Excel (version 2010) for further analysis. Data exported to Excel were edited, sorted, and categorized based on inclusion and exclusion criteria, year of publications, countries, regions, and fields. The data set was later imported to the VOSviewer and used to create network maps of co-authorships of authors and countries, co-occurrence of indexed keywords, citation of authors, citations of countries, bibliographic coupling of authors and countries, and co-citations of cited source with authors.

RESULTS AND DISCUSSION

A previous systematic study demonstrated that the analysis of scientific yield from their citations requires the execution of laboratory research to commercialization. Moreover, collective observation indicated that the knowledge of researchers about published scientific documents is either partial or inadequate due to the non-systemic presentation of research data and publication amid various scientific databases, which make it difficult to interpret available documents. Bibliometric analysis solves the above-mentioned concerns by systemized incorporation and presentation of all kinds of research executed in a certain field [89, 90]. The results of the bibliometric analysis of data obtained from 2000-2020 on the Scopus database were analyzed using VOSviewer. The analysis was based on the year of publication, countries, journal participation, co-authorship of countries, co-occurrence of authors, keywords, and bibliographies. Analysis parameters were chosen to highlight key carters of the research relating to bio-based pharmaceutical excipients. This study is aimed to help identify eminent works, innovations, and trends concerning bio-based pharmaceutical excipients.

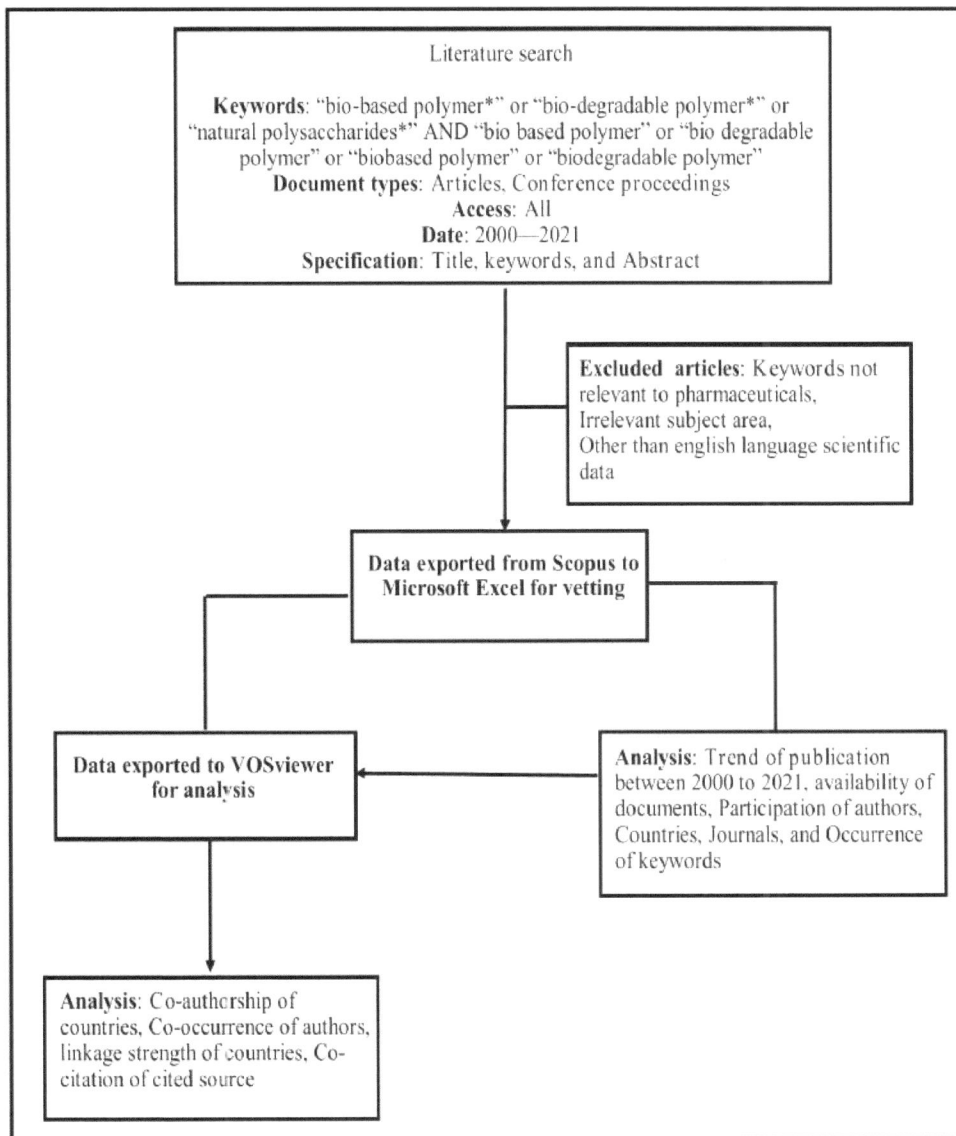

Fig. (1). Process flowchart for meta-analysis of data retrieved from Scopus database.

Publication Trends During 2000 to 2020 and Distribution of Publication Based on the Theme

A total of 2923 research articles and conference proceedings on bio-based polymer were exported from the Scopus database into VOSviewer software. Fig. (**2**) indicates the yearly trends of publication with citations for retrieved data.

From the data exported, research articles constituted 97.7% and proceedings were 2.3%. As presented in Fig. (**2a**), the trend of publication for research articles and proceedings gradually increased from 2000 to 2010, however, it declined in 2011. A drastic increase in published research on bio-based polymers, with variability in the total number of publications in 2016 was observed. Moreover, 72.06% of the retrieved documents were published within the last 10 years. This might probably be due to factors such as the ease of availability, multifunctional utility, public perception, and an increase in demand for biodegradable natural polymers. The variation in the percentage of published documents with citations indicated the growing interest of researchers in research on biomaterials. Considering the overall scenario and trends, and the times span it takes to pass a chapter from writing to publication, it is likely going to be a temporary increase in the number of publications. The retrieved documents were further analyzed based on subjects with citations. It was observed that natural polymer theme-based research was conducted by scientists of various fields. The results revealed that chemistry (37%) based research had the highest percentage of contribution in scientific data followed by material science and engineers (31%) and chemicals with pharmaceutical (25%).

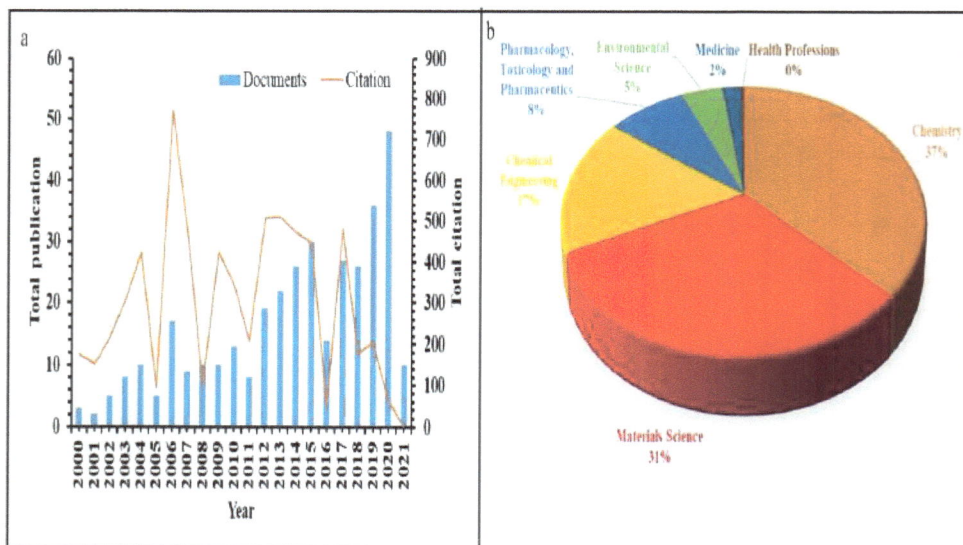

Fig. (2). A total number of publications with citations for bio-based polyssacrides research during 2000-2021 (**a**) and documents by subject area (**b**). (For interpretation of the results to color in this Fig. legend, the reader referred to either web version of this chapter or color print).

Co-Authorship of Authors

The co-author's network analysis was conducted to determine the average publications on pharmaceutical natural excipients using network and overlay visualization of the VOSviewer software. A high number of citations for research documents indicate researcher expertise as well as the visibility of their research. A total of 1397 authors were recorded in co-authorship analysis. Out of which, only 168 authors had a minimum of 2 publications (articles and conference proceedings research papers) on pharmaceuticals in bio-based and related fields from 2000 to 2020 as presented in Fig. (**3**). The author Liu Y. had the highest number of published papers (10 documents), followed by Singh B (8 documents), and Xu W (7 documents) (Table **2**). Furthermore, the analysis indicated the total number of citations and link strength of 193, 324, 234, and 30, 18, 19 for Liu, Singh, and Xu respectively. Moreover, the author's number of publications is also used as an indicator of productivity in a particular research field. In addition, the total link strength indicates partnership among authors. Recently a trend of self-citations where an author gives citations to himself reflects a discrepancy of an author's achievements. This trend is likely one of the major causes of inconsistency during the setting of criteria for ranking of authors, journals, institutions, and countries [91]. While various researchers believe that citation numbers are a reflection of the significance [92] and some others contemplate that the number of publications is an indication of the author's productivity in the particular research field [93].

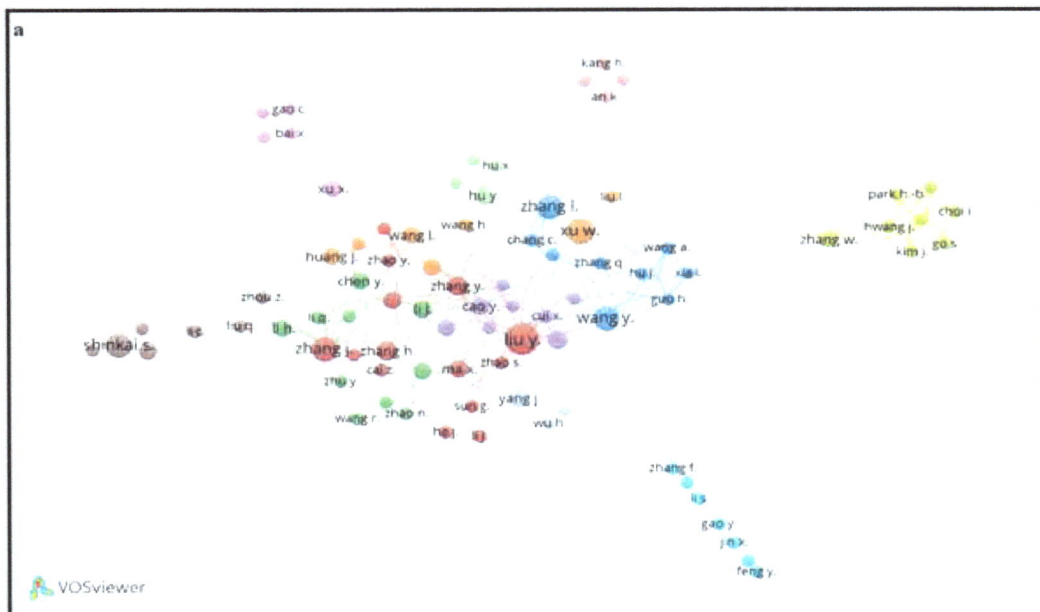

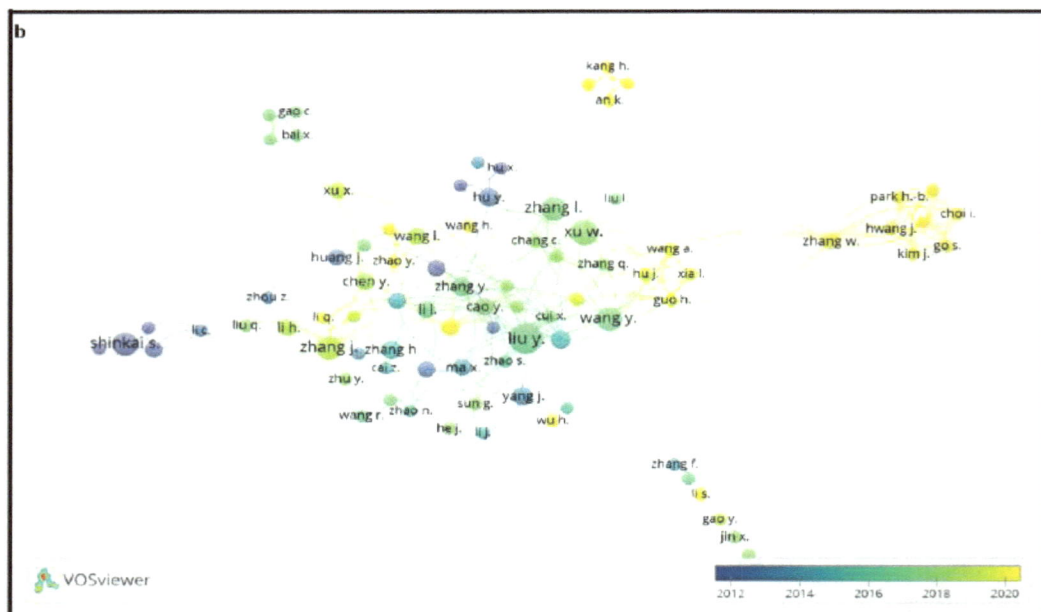

Fig. (3). Network **(a)** and overlay **(b)** visualization of author's co-authorships with a minimum of 2 publications from 2000 – 2020. This typical constellation is indicative of publication with similar keywords in the author's publications. In addition, collated author used most commonly "bio-based" excipients as a keyword. (For interpretation of the results to color in this Fig. legend, the reader is referred to either web version of this chapter or color print).

Table 2. Bibliometric analysis of the highest cited authors documents with the keyword of "bio-based polysaccharides*" based on a minimum number of cited documents of an author (02) with total linkage strength out of 1397 authors.

Authors	Documents	Citations	Total Link Strength	Authors	Documents	Citations	Total Link Strength
Singh B.	8	324	33	Polyakov N.E.	4	84	21
Chauhan N.	6	269	24	Li c.	2	80	2
Shinkai S.	6	261	8	Zhang L.	6	80	7
Xu W.	7	233	0	Yang J.	4	75	0
Chauhan G.S.	4	213	27	Kumar K.	3	74	22
Liu Y.	10	193	0	Cai Z.	2	72	3
Zhang H.	4	191	3	Liu H.	3	72	0
Huang J.	3	172	0	Park S.	3	67	10
Le Cloirec P.	4	170	0	Andres Y.	2	66	0
Reddad Z.	4	170	0	Dushkin A.Y.	2	66	13
Sakurai K.	2	154	4	Gerente C.	2	66	0

Authors	Documents	Citations	Total Link Strength	Authors	Documents	Citations	Total Link Strength
Ma X.	3	144	0	Kispert L.D.	2	66	13
Wang W.	3	138	0	Konovalova T.A.	2	66	13
Wang Y.	6	137	3	Leshina T.V.	2	66	13
Dellacherie E.	2	136	0	Meteleva E.S.	2	66	13
Kumar S.	3	132	8	Chang C.	2	62	7
Wang C.	4	127	0	Ye D.	2	62	7
Andrès Y.	2	104	0	Sharma N.	3	59	11
Gérente C.	2	104	0	Zhou Z.	2	58	0
Numata M.	3	103	2	Zhang Q.	2	56	3
Mishra A.	6	97	27	Lee S.Y.	2	52	8
Zhang F.	2	96	0	Li L.	3	52	3
Haraguchi S.	2	94	6	Srinivasan R.	3	52	22
Sharma D.K.	3	90	8	Hu Y.	4	50	0

Occurrence of Authors Keywords and Co-Occurrence of all Keywords

The author's keywords were retrieved and analyzed. Similar keywords were used by authors. A total of 986 keywords were retrieved among which 156 met the threshold requirements of the minimum number of two times occurrence for a keyword. The most commonly used keywords were polysaccharides, chitosan, alginate, adsorption, hydrogels, and Phyllium. Polysaccharides had a total occurrence of 26 with total linkage strength of 50 whereas, chitosan and alginate had an occurrence of 47 and 25. The retrieved data indicated the presence of a total of 16 clusters in the keywords used by authors. The overlay visualization of keywords (Fig. **4**) demonstrated a clear trend in a shift of research awareness of novel bio-based pharmaceutical excipients with intellectual property rights and approval from food and drug administrations. Furthermore, the exploration revealed that keywords such as polysaccharides, chitosan, alginate, adsorption, hydrogels, psyllium, pectin, biomaterials, and biocompatibility were popularly used till 2015. However, recently emphasized keywords used are tissue engineering, hybrid material, antioxidant capacity, 3D printing, dye removal, and amylose.

Author's keywords in the field of pharmaceutical excipients were retrieved and analyzed using VOSviewer algorithm mapping techniques. Keywords in a scientific publication are of great importance in knowing the research trends and follow-up as well as identifying the research gap. A total of 3685 keywords with

962 co-occurrences were obtained with a minimum threshold occurrence of 5 during the last 20 years. The co-occurring keys were grouped in 5 colored basic cultures depending on the similarities in the study. The 5 most occurring keywords were "natural polysaccharide" (167 occurrences), "polysaccharides" (103 occurrences), "polysaccharide" (74 occurrences), "article" (82 occurrences), and "hydrogel" (53 occurrences). In addition, the topmost occurring keywords demonstrated a total link strength of 1206, 828, 966, 770, and 546 for natural polysaccharides, polysaccharides, polysaccharides, articles, and hydrogel respectively. The study revealed that there is a global awareness of the utility of biomaterials as pharmaceutical excipients. Moreover, the results indicated that most of the research was focused on natural polysaccharides isolation, structural analysis, and pharmaceutical application with functional modification. A similar bibliometric analysis reported that the total number of articles retrieved for study using either Web of Science or the Scopus may not represent 100% accurate data on bio-based pharmaceutical excipients since several journals are not indexed [94]. Although the authors confirmed all terms used were relevant to the search study, still there is the possibility of scarce false-positive results. Further, errors can ascend from spelling differences that might occur due to the recurrence of an item as seen in some of the network maps acquired [95]. Fig. (**5**) shows the overlay and density visualization of co-occurrence of the selected keywords with boolean terminology "AND" "OR" used by the researcher in a publication related to bio-based polymers.

Fig. (4). Overlay visualization of author's keywords co-occurrence. (For interpretation of the results to color in this Fig. legend, the reader is referred to either web version of this chapter or color print).

Fig. (5). Density **(a)** and overlay **(b)** visualization of all keywords for co-occurrence with a minimum threshold occurrence of 5 during the last 20 years. (For interpretation of the results to color in this Fig. legend, the reader is referred to either web version of this chapter or color print).

Bibliographic Coupling of Countries

Bibliographic coupling of authors between countries helps to understand the international research collaboration in specific research fields. Data were analyzed at a maximum of at least 2 articles published or documents per country. The results indicated that 59 countries fitted into the set parameter with a threshold of 16. The bibliographic coupling of countries indicated 8 clusters with 34 linkage and overall linkage strength of 6885. The network visualization (Fig. **6**) showed that India had the highest number of documents [83], with 1691 citations and total linkage strength of 7 followed by China, United States, France, Japan, and the United Kingdom with citations and linkage strength of 1362, 681, 517, 431, 409 and 6, 6, 0, 3 respectively. The output of retrieved data suggested that these countries led the research in the development of biomaterials. The relative strength of collaborative coupling was measured by the thickness of linking lines between countries [96]. The results indicated that India, China, the United States of America, France, and the United Kingdom are currently involved in research relating to bio-based excipients. Although countries such as France, Brazil, and Spain were involved in the current topic of research as reflected by their number of published documents with citations, international collaboration was not encouraging. Moreover, the results indicated poor international collaboration of Southeast Asian and African with the international community. Bibliometric analysis of the keyword "Bio-based polysaccharides*" based on a minimum number of cited documents of a country (05) with total linkage strength is presented in Table **3**.

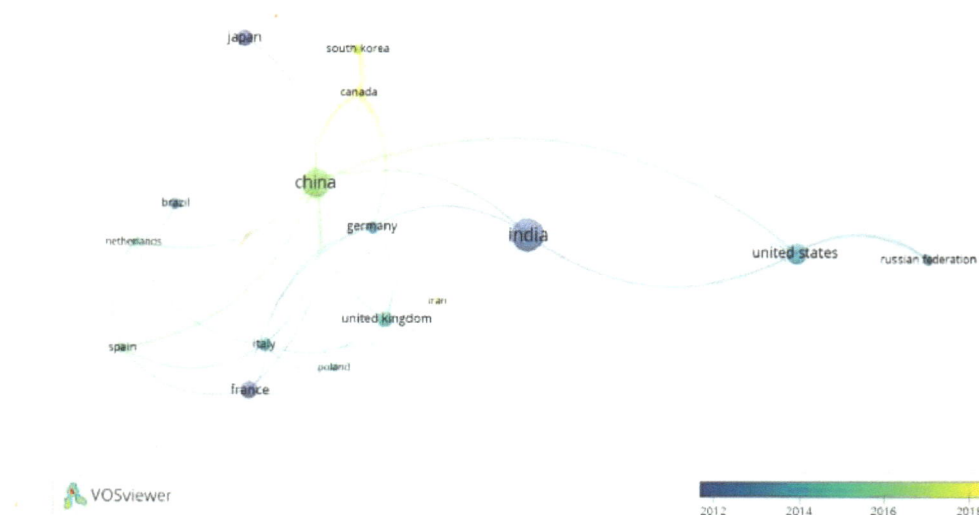

Fig. (6). Network visualization of bibliographic coupling of countries. (For interpretation of the results to color in this Fig. legend, the reader is referred to either web version of this chapter or color print).

Table 3. Bibliometric analysis of keyword "bio-based polysaccharides*" based on the minimum number of cited documents of a country (05) with total linkage strength.

Country	Documents	Citations	Total Link Strength
India	83	1691	7
China	78	1362	6
Russian federation	26	164	3
Japan	25	431	3
United States	24	681	6
Italy	20	284	2
France	16	517	0
Brazil	12	210	0
United kingdom	10	409	1
Canada	9	147	1
Spain	9	139	0
South Korea	8	136	1
Poland	8	90	0
Iran	8	62	0
Germany	7	258	0
Netherlands	5	81	0

Co-Citation of Sources

The acquired data were analyzed based on journal participation and publication in the field of bio-based polymers. A total of 180 source journals were found to have published at least one article on the topic. Among the journals, 50 had at least 2 documents with a minimum of one citation. The top 5 active journals including Carbohydrate Polymer, Environmental Science and Technology, International Journal of Pharmaceutics, Journal of Colloids and Interface, and Journal of Applied Polymer Science accounted for 26.9, 8.05, 7.95, 6.0, and 5.73% of total documents citations, respectively. Stratification of documents based on pharmaceutical excipients classification of journals indicated that 60.74% of retrieved research documents were published by sources related to polysaccharides, while 39.26% were attributed to the allied pharmaceutical field (Fig. 7).

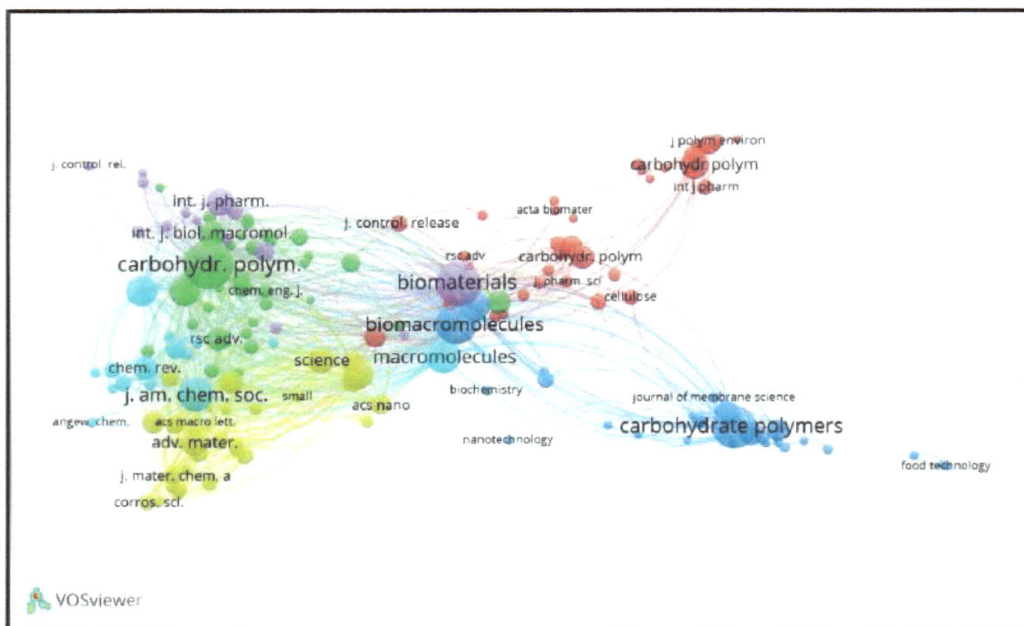

Fig. (7). Network visualization for co-citation of cited sources. (For interpretation of the results to color in this Fig. legend, the reader is referred to either web version of this chapter or color print).

Limitations of the Study

The current bibliometric analysis presents research trends in the field of bio-based pharmaceutical excipients for various applications. Although, the authors tried to ensure that the information presented represents the obtained trends from the Scopus database and in addition, scrutinized the documents to avoid duplications which can cause errors in either network visualization demography images using VOSveiwer software [97]. However, this data might still be incomplete since some of the high-impact research documents published from 2000 to 2020 may not be included in the Scopus database and could not able to attain the focus of potential researchers who might have transferred the technology from pilot scale to large scale. Moreover, there might also be other potential influences that might affect the current analysis such as the use of a single electronic database that is Scopus, to conduct the study. Hence, a reader must keep in mind that the retrieved data results might significantly differ if we have used Google Scholar or any other database [98].

CONCLUSIONS

A total of 2923 research and conference proceedings analyzed, indicated rising trends in cumulative publication and citations reflecting a nascent interest in the field of bio-based pharmaceutical excipients. Analyzed data indicated that several other research works in allied pharmaceutical fields were also carried out during 2000 to 2020. Researchers have also shown interest in tissue engineering and nanocomposite fabrication using biomaterials. The structural modification of biomaterials was also an area of interest between 2010-2016. Journals with a higher number of research in this field include Carbohydrate Polymers, Biomacromolecules, and Biomaterials accompanied by high numbers of citations. The current chapter presents the bibliometric analysis of published articles related to bio-based excipients for a period of 20 years (2000-2020), to understand the research status and trends related to the field of study as well as the collaborative efforts of researchers. Moreover, both qualitative and quantitive analyses show the appreciation of major advances and progress in the area of biomaterials required to accelerate the progress of the study, and reveal the trends of research and development.

REFERENCES

[1]   Yu X, Buevich AV, Li M, Wang X, Rustum AM. A compatibility study of a secondary amine active pharmaceutical ingredient with starch: identification of a novel degradant formed between desloratadine and a starch impurity using LC-MS(n) and NMR spectroscopy. J Pharm Sci 2013; 102(2): 717-31.
        [http://dx.doi.org/10.1002/jps.23416] [PMID: 23242759]

[2]   Wu Y, Levons J, Narang AS, Raghavan K, Rao VM. Reactive impurities in excipients: profiling, identification and mitigation of drug-excipient incompatibility. AAPS PharmSciTech 2011; 12(4): 1248-63.
        [http://dx.doi.org/10.1208/s12249-011-9677-z] [PMID: 21948318]

[3]   William BJ. Ask the historian: The origin of polymer concept J Chem Edu. 2008; 88: 624-5.

[4]   Smeds KA, Grinstaff MW. Synthesis of photocrosslinkable biopolymers for *in situ* applications. Proceedings of American Chemical Society, Polymer Preprints 2000; 1722-3.

[5]   Chan LW, Lee HY, Heng PWS. Production of alginate microspheres by internal gelation using an emulsification method. Int J Pharm 2002; 242(1-2): 259-62.
        [http://dx.doi.org/10.1016/S0378-5173(02)00170-9] [PMID: 12176259]

[6]   Jiang Z, Zhang X, Wu L, *et al.* Exolytic products of alginate by the immobilized alginate lyase confer antioxidant and antiapoptotic bioactivities in human umbilical vein endothelial cells. Carbohydr Polym 2021; 251: 116976.
        [http://dx.doi.org/10.1016/j.carbpol.2020.116976] [PMID: 33142553]

[7]   Yokosawa MM, Frollini E. Effect of the addition of a cationic derivative of the natural polysaccharide guar gum on the stability of an aqueous dispersion of alumina. J Macromol Sci Pure Appl Chem 2002; A(7): 709-21.

[8]   El Tahlawy K, Hudson SM. Synthesis of a well-defined chitosan graft poly(methoxy polyethyleneglycol methacrylate) by atom transfer radical polymerization. J Appl Polym Sci 2003; 89(4): 901-12.
        [http://dx.doi.org/10.1002/app.12001]

[9]   Mishra A, Agarwal M, Yadav A. Fenugreek mucilage as a flocculating agent for sewage treatment. Colloid Polym Sci 2003; 281(2): 164-7.
        [http://dx.doi.org/10.1007/s00396-002-0765-1]

[10]   Mishra A, Srinivasan R, Gupta R. P. psyllium-g-polyacrylonitrile: synthesis and characterization. Colloid Polym Sci 2003; 281(2): 187-9.
[http://dx.doi.org/10.1007/s00396-002-0777-x]

[11]   Mishra A, Yadav A, Agarwal M, Srinivasan R. Plantago psyllium-grafted-polyacrylonitrile: Synthesis, characterization and its use for solid removal from sewage wastewater. Chin J Polym Sci 2005; 23(1): 113-8. [Engl Ed].
[http://dx.doi.org/10.1142/S0256767905000072]

[12]   Murata Y, Tsumoto K, Kofuji K, Kawashima S. Effects of natural polysaccharide addition on drug release from calcium-induced alginate gel beads. Chem Pharm Bull (Tokyo) 2003; 51(2): 218-20.
[http://dx.doi.org/10.1248/cpb.51.218] [PMID: 12576662]

[13]   Satoh T, Nishiyama K, Nagahata M, Teramoto A, Abe K. The research on physiological property of functionalized hyaluronan: interaction between sulfated hyaluronan and plasma proteins. Polym Adv Technol 2004; 15(12): 720-5.
[http://dx.doi.org/10.1002/pat.486]

[14]   Yang J, Du Y, Huang R, *et al.* Chemical modification and antitumour activity of Chinese lacquer polysaccharide from lac tree *Rhus vernicifera.* Carbohydr Polym 2005; 59(1): 101-7.
[http://dx.doi.org/10.1016/j.carbpol.2004.09.004]

[15]   Zhang C, Huang K. Characteristic immunostimulation by MAP, a polysaccharide isolated from the mucus of the loach, *Misgurnus anguillicaudatus.* Carbohydr Polym 2005; 59(1): 75-82.
[http://dx.doi.org/10.1016/j.carbpol.2004.08.023]

[16]   Carr LG, Parra DF, Ponce P, Lugão AB, Buchler PM. Influence of fibers on the mechanical properties of cassava starch foams. J Polym Environ 2006; 14(2): 179-83.
[http://dx.doi.org/10.1007/s10924-006-0008-5]

[17]   Kupchik LA, Kartel' NT, Bogdanov ES, Bogdanova OV, Kupchik MP. Chemical modification of pectin to improve its sorption properties. Russ J Appl Chem 2006; 79(3): 457-60.
[http://dx.doi.org/10.1134/S1070427206030256]

[18]   Zimin YS, Borisova NS, Kutlugildina GG, Mudarisova RK, Borisov IM, Mustafin AG. Oxidation and destruction of arabinogalactan and pectins under the action of hydrogen peroxide and ozone-oxygen mixture. React Kinet Mech Catal 2017; 120(2): 673-90.
[http://dx.doi.org/10.1007/s11144-016-1113-7]

[19]   Mishra A, Bajpai M. Removal of sulphate and phosphate from aqueous solutions using a food grade polysaccharide as flocculant. Colloid Polym Sci 2006; 284(4): 443-8.
[http://dx.doi.org/10.1007/s00396-005-1399-x]

[20]   Malviya R, Srivastava P, Kumar U, Bhargava CS, Sharma PK. Formulation and comparison of suspending properties of different natural polymers using paracetamol suspension. Int J Drug Dev Res 2010; 2(4): 886-91.

[21]   Sravani B, Deveswaran R, Bharath S, Basavaraj BV, Madhavan V. Development of sustained release metformin hydrochloride tablets using a natural polysaccharide. Int J Appl Pharm 2012; 4(2): 23-9.

[22]   Kaur G, Mahajan M, Bassi P. Derivatized polysaccharides: Preparation, characterization, and application as bioadhesive polymer for drug delivery. Int J Polym Mater 2013; 62(9): 475-81.
[http://dx.doi.org/10.1080/00914037.2012.734348]

[23]   Radha GV, Santosh Naidu M. Design of transdermal films of alfuzocin HCl by using a natural polymer tamarind seed polysaccharide extract. Pharm Lett 2013; 5(3): 457-64.

[24]   Sahu VK, Sharma N, Sahu PK, Saraf SA. Formulation and evaluation of floating-mucoadhesive microspheres of novel natural polysaccharide for site specific delivery of ranitidine hydrochloride. International Journal of Applied Pharmaceutics 2017; 9(3): 15-9.
[http://dx.doi.org/10.22159/ijap.2017v9i3.16137]

[25]   Mishra A, Bajpai M, Pandey S. Removal of dyes by biodegradable flocculants: A lab scale investigation. Sep Sci Technol 2006; 41(3): 583-93.
[http://dx.doi.org/10.1080/01496390500526110]

[26]   Olas B, Saluk-Juszczak J, Pawlaczyk I, *et al.* Antioxidant and antiaggregatory effects of an extract from *Conyza canadensis* on blood platelets *in vitro*. Platelets 2006; 17(6): 354-60.
[http://dx.doi.org/10 1080/09537100600746805] [PMID: 16973495]

[27]   Singh B, Chauhan GS, Bhatt SS, Kumar K. Metal ion sorption and swelling studies of psyllium and acrylic acid based hydrogels. Carbohydr Polym 2006; 64(1): 50-6.
[http://dx.doi.org/10.1016/j.carbpol.2005.10.022]

[28]   Rathnanand M, Narkhede R, Udupa N, Kalra A. Development of novel floating delivery system based on psyllium: application on metformin hydrochloride. Curr Drug Deliv 2013; 10(3): 336-42.
[http://dx.doi.org/10.2174/1567201811310030010] [PMID: 23410070]

[29]   Singh B, Chauhan GS, Kumar S, Chauhan N. Synthesis, characterization and swelling responses of pH sensitive psyllium and polyacrylamide based hydrogels for the use in drug delivery (I). Carbohydr Polym 2007; 67(2): 190-200.
[http://dx.doi.org/10.1016/j.carbpol.2006.05.006]

[30]   Lu D, Wen X, Liang J, Zhang X, Gu Z, Fan Y. Novel pH-sensitive drug delivery system based on natural polysaccharide for doxorubicin release. Chin J Polym Sci 2008; 26(3): 369-74.
[http://dx.doi.org/10.1142/S0256767908003023]

[31]   Srinivasan R, Mishra A. Okra (*Hibiscus esculentus*) and Fenugreek (*Trigonella foenum* graceum) mucilage: Characterization and application as flocculants for textile effluent treatments. Chin J Polym Sci 2008; 26(6): 679-87.
[http://dx.doi.org/10.1142/S0256767908003424]

[32]   Gong Y, Wang C, Lai RC, Su K, Zhang F, Wang D. An improved injectable polysaccharide hydrogel: modified gellan gum for long-term cartilage regeneration *in vitro*. J Mater Chem 2009; 19(14): 1968-77.
[http://dx.doi.org/10.1039/b818090c]

[33]   Adrover A, Paolicelli P, Petralito S, *et al.* Gellan gum/laponite beads for the modified release of drugs: experimental and modeling study of gastrointestinal release. Pharmaceutics 2019; 11(4): 187.
[http://dx.doi.org/10.3390/pharmaceutics11040187] [PMID: 30999609]

[34]   Singh S, Ushir Y, Chidrawar RV, Vadalia KR, Sheth NR, Singh S. Preliminary evaluation of *Cassia auriculata* seed mucilage as binding agent. Pharmacogn J 2009; 1(4): 251-7.

[35]   Singh S, Gendle R, Sheth N, Patel RD. Isolation and evaluation of binding property of *Zizyphus jujuba* lamk. Seed mucilage in tablet formulations. J Glob Pharma Technol 2010; 2(1): 98-102.

[36]   Singh S, Bothara SB, Singh S, Patel RD, Mahobia NK. Pharmaceutical characterization of *Cassia tora* of seed mucilage in tablet formulations. Pharm Lett 2010; 2(5): 54-61.

[37]   Singh S, Sangeeta S. Preliminary investigation of *Cassia sophera* linn seed mucilage in tablet formulations. Int J Pharm Appl Sci 2010; 1(1): 63-9.

[38]   Singh S, Sheth NR, Singh S, Rajesh D. Pharmaceutical characterization of *Prosopis juliflora* (Sw) seed mucilage-excipient. Acta Pharmaceutica Sciencia 2010; 52(4): 487-94.

[39]   Singh S, Sangeeta S. Isolation and evaluation of *Cassia aungostifolia* seed mucilage as granulating agent. Int J Pharm Sci Res 2010; 1(2): 118-25.

[40]   Singh S, Sangeeta S, Bothara SB, Patel RD. Pharmaceutical characterization of some natural excipient as potential mucoadhesives agent. Pharm Res 2010; 4: 91-104.

[41]   Singh S, Sangeeta S. Evaluation of *Cassia fistula* Linn. seed mucilage in tablet formulations. Int J Pharm Res 2010; 2(3): 1839-46.

[42]   Bothara SB, Singh S. Thermal studies on natural polysaccharide. Asian Pac J Trop Biomed 2012; 2(2): S1031-5.
        [http://dx.doi.org/10.1016/S2221-1691(12)60356-6]

[43]   Singh S, Bothara SB. Formulation, development of oral mucoadhesive tablets using natural materials extracted from indgenously edible fruits available in chhattisgrah. Inventi Rapid: Novel Excipients

[44]   Singh S, Bothara SB. Formulation development of oral mucoadhesive tablets of losartan potassium using mucilage isolated from *Diospyros melonoxylon* Roxb Seeds. International Journal of Pharmaceutical Sciences and Nanotechnology 2013; 6(3): 2154-63.
        [http://dx.doi.org/10.37285/ijpsn.2013.6.3.7]

[45]   Singh S, Bothara SB. Development of oral mucoadhesive tablets of losartan potassium using natural gum from *Manilkara Zapota* seeds. International Journal of Pharmaceutical Sciences and Nanotechnology 2013; 6(4): 2245-54.
        [http://dx.doi.org/10.37285/ijpsn.2013.6.4.7]

[46]   Singh S, Ahmad A, Bothara B S. Formulation of oral mucoadhesive tablets using mucilage isolated from *Buchanania lanzan* spreng seeds. International Journal of Pharmaceutical Sciences and Nanotechnology 2014; 7(2): 2494-503.
        [http://dx.doi.org/10.37285/ijpsn.2014.7.2.11]

[47]   Singh S, Bothara SB. Morphological, physico-chemical and structural characterization of mucilage isolated from the seeds of *Buchanania lanzan* Spreng. Int J Health Allied Sci 3(1): 33-5.

[48]   Singh S, Bothara SB. Physico-chemical and structural characterization of mucilage isolated from seeds of *Diospyros melonoxylon* Roxb. Braz J Pharm Sci 2014; 50(4): 713-25.
        [http://dx.doi.org/10.1590/S1984-82502014000400006]

[49]   Singh S, Sunil B B. *In vivo* mucoadhesive strength appraisal of gum *Manilkara zapota*. Braz J Pharm Sci 2015; 51(3): 689-98.
        [http://dx.doi.org/10.1590/S1984-82502015000300021]

[50]   Singh S, Nwabor OF, Ontong JC, Voravuthikunchai SP. Characterization and assessment of compression and compactibility of novel spray-dried, co-processed bio-based polymer. J Drug Deliv Sci Technol 2020; 56: 101526.
        [http://dx.doi.org/10.1016/j.jddst.2020.101526]

[51]   Singh S, Nwabor OF, Ontong JC, Kaewnopparat N, Voravuthikunchai SP. Characterization of a novel, co-processed bio-based polymer, and its effect on mucoadhesive strength. Int J Biol Macromol 2020; 145: 865-75.
        [http://dx.doi.org/10.1016/j.ijbiomac.2019.11.198] [PMID: 31783076]

[52]   Maru SG, Singh S. Physicochemical and mucoadhesive strength characterization of natural polymer obtained from leaves of *Aloe vera*. Pharmtechmedica 2013; 2(3): 303-8.

[53]   Singh S, Maru SG, Bothara B S. Formulation and *ex-vivo* evaluation of glipizide buccal tablets. International Journal of Pharmaceutical Sciences and Nanotechnology 2014; 7(2): 2487-93.
        [http://dx.doi.org/10.37285/ijpsn.2014.7.2.10]

[54]   Singh S, Goswami H. Formulation of oral mucoadhesive tablets of pioglitazone using natural gum from seeds of *Pithecellobium dulce*. International Journal of Pharmaceutical Sciences and Nanotechnology 2015; 8(4): 3031-8.
        [http://dx.doi.org/10.37285/ijpsn.2015.8.4.6]

[55]   Singh S, Santoki R. Development and appraisal of mucoadhesive tablets of hydralazine using isolated mucilage of *Annona squamosa* seeds. Int J Pharm Sci Nanotech 2016; 9(4): 3349-56.

[56]   Wang LF, Duan JC, Miao WH, Zhang RJ, Pan SY, Xu XY. Adsorption–desorption properties and characterization of crosslinked Konjac glucomannan-graft-polyacrylamide-co-sodium xanthate. J Hazard Mater 2011; 186(2-3): 1681-6.
        [http://dx.doi.org/10.1016/j.jhazmat.2010.12.055] [PMID: 21236570]

[57]   Jayakumar GC, Kanth SV, Sai KP, Chandrasekaran B, Rao JR, Nair BU. Scleraldehyde as a stabilizing agent for collagen scaffold preparation. Carbohydr Polym 2012; 87(2): 1482-9.
[http://dx.doi.org/10.1016/j.carbpol.2011.09.042]

[58]   Chao Z, Ri-fu Y, Tai-qiu Q. Ultrasound-enhanced subcritical water extraction of polysaccharides from *Lycium barbarum* L. Separ Purif Tech 2013; 120: 141-7.
[http://dx.doi.org/10.1016/j.seppur.2013.09.044]

[59]   Saikia P, Sahu BP, Dash SK. Isolation and characterisation of some natural polysaccharides as pharmaceutical excipients. Int J Pharm Res 2013; 5(3): 1196-206.

[60]   Sharma HK, Lahkar S, Kanta Nath L. Formulation and *in vitro* evaluation of metformin hydrochloride loaded microspheres prepared with polysaccharide extracted from natural sources. Acta Pharm 2013; 63(2): 209-22.
[http://dx.doi.org/10.2478/acph-2013-0019] [PMID: 23846143]

[61]   Godge GR, Hiremath SN. Development and evaluation of colon targeted drug delivery system by using natural Polysaccharides/Polymers. Dhaka University Journal of Pharmaceutical Sciences 2015; 13(1): 105-13.
[http://dx.doi.org/10.3329/dujps.v13i1.21874]

[62]   Newton AMJ, Swathi P, Kumar N, Manoj Kumar K. A comparative study of Okra gum on controlled release kinetics and other formulation characteristics of tramadol HCL extended release matrix tablets vs synthetic hydrophilic polymers. Int J Drug Deliv 2014; 6(4): 339-50.

[63]   Norström E, Fogelström L, Nordqvist P, Khabbaz F, Malmström E. Gum dispersions as environmentally friendly wood adhesives. Ind Crops Prod 2014; 52: 736-44.
[http://dx.doi.org/10.1016/j.indcrop.2013.12.001]

[64]   Sharma VK, Mazumdar B. Characterization of gliclazide release from Isabgol husk hydrogel beads by validated HPLC method. Acta Pol Pharm 2014; 71(1): 153-66.
[PMID: 24779204]

[65]   Heidarieh M, Shawrang P, Akbari M, Heidarieh H. Proximate analysis of different groups of irradiated alginic acid. Intl J Radiat Res 2015; 13(2): 187-90.

[66]   Nair RS, Reshmi T, Reshmi P, Sarika C, Snima KS, Akk U, *et al.* Biocompatibility studies of pectin-fibrin nanocomposite bearing BALB/c mice. Cellul Chem Technol 2015; 49(1): 55-60.

[67]   Devendiran RM, Chinnaiyan S, Yadav NK, *et al.* Green synthesis of folic acid-conjugated gold nanoparticles with pectin as reducing/stabilizing agent for cancer theranostics. RSC Advances 2016; 6(35): 29757-68.
[http://dx.doi.org/10.1039/C6RA01698G]

[68]   Zhao S, Zhang Y, Liu Y, *et al.* Preparation and optimization of calcium pectate beads for cell encapsulation. J Appl Polym Sci 2018; 135(2): 45685.
[http://dx.doi.org/10.1002/app.45685]

[69]   Soares PAG, C de Seixas JRP, Albuquerque PBS, *et al.* Development and characterization of a new hydrogel based on galactomannan and κ-carrageenan. Carbohydr Polym 2015; 134: 673-9.
[http://dx.doi.org/10.1016/j.carbpol.2015.08.042] [PMID: 26428171]

[70]   Ström A, Larsson A, Okay O. Preparation and physical properties of hyaluronic acid□based cryogels. J Appl Polym Sci 2015; 132(29): app.42194.
[http://dx.doi.org/10.1002/app.42194]

[71]   Anuradha K, Bangal P, Madhavendra SS. Macromolecular arabinogalactan polysaccharide mediated synthesis of silver nanoparticles, characterization and evaluation. Macromol Res 2016; 24(2): 152-62.
[http://dx.doi.org/10.1007/s13233-016-4018-4]

[72]   Hosseini SF, Rezaei M, Zandi M, Farahmandghavi F. Preparation and characterization of chitosan nanoparticles-loaded fish gelatin-based edible films. J Food Process Eng 2016; 39(5): 521-30.

[http://dx.doi.org/10.1111/jfpe.12246]

[73]   Lodeiro P, Achterberg EP, Pampín J, Affatati A, El-Shahawi MS. Silver nanoparticles coated with natural polysaccharides as models to study AgNP aggregation kinetics using UV-Visible spectrophotometry upon discharge in complex environments. Sci Total Environ 2016; 539: 7-16.
[http://dx.doi.org/10.1016/j.scitotenv.2015.08.115] [PMID: 26363390]

[74]   Winkworth-Smith CG, MacNaughtan W, Foster TJ. Polysaccharide structures and interactions in a lithium chloride/urea/water solvent. Carbohydr Polym 2016; 149: 231-41.
[http://dx.doi.org/10.1016/j.carbpol.2016.04.102] [PMID: 27261747]

[75]   Chan SY, Chan BQY, Liu Z, *et al.* Electrospun pectin-polyhydroxybutyrate nanofibers for retinal tissue engineering. ACS Omega 2017; 2(12): 8959-68.
[http://dx.doi.org/10.1021/acsomega.7b01604] [PMID: 30023596]

[76]   Xie C, Feng Y, Cao W, Xia Y, Lu Z. Novel biodegradable flocculating agents prepared by phosphate modification of Konjac. Carbohydr Polym 2007; 67(4): 566-71.
[http://dx.doi.org/10.1016/j.carbpol.2006.06.036]

[77]   Ma Y, Lei Y, Zhang X, Liu Y, Sui H, Feng J, *et al.* Konjac glucomannan and its modifications as tablet disintegrants. Acta Pol Pharm Drug Res 2017; 74(6): 1841-9.

[78]   An K, Kang H, Zhang L, Guan L, Tian D. Preparation and properties of thermosensitive molecularly imprinted polymer based on konjac glucomannan and its controlled recognition and delivery of 5-fluorouracil. J Drug Deliv Sci Tech 2020; p. 60.

[79]   Bashir S, Teo YY, Ramesh S, Ramesh K. Synthesis and characterization of karaya gum-g- poly (acrylic acid) hydrogels and *in vitro* release of hydrophobic quercetin. Polymer (Guildf) 2018; 147: 108-20.
[http://dx.doi.org/10.1016/j.polymer.2018.05.071]

[80]   Chesnokova NY, Levochkina LV, Prikhod'ko YV, Kuznetsova AA, Vladykina TV. Influence of polysaccharide functional groups on the extraction degree of blackcurrant anthocyanin pigment. J Pharm Sci Res 2018; 10(3): 659-61.

[81]   Jalali S, Montazer M, Malek RMA. A novel semi-bionanofibers through introducing tragacanth gum into pet attaining rapid wetting and degradation. Fibers Polym 2018; 19(10): 2088-96.
[http://dx.doi.org/10.1007/s12221-018-8276-y]

[82]   Kumar V, Rehani V, Kaith BS, Saruchi S. Synthesis of a biodegradable interpenetrating polymer network of Av-cl-poly(AA-ipn-AAm) for malachite green dye removal: kinetics and thermodynamic studies. RSC Advances 2018; 8(73): 41920-37.
[http://dx.doi.org/10.1039/C8RA07759B] [PMID: 35558783]

[83]   Zhuang Y, Wang L, Liu C, *et al.* A novel fiber from Bletilla striata tuber: physical properties and application. Cellulose 2019; 26(9): 5201-10.
[http://dx.doi.org/10.1007/s10570-019-02472-3]

[84]   Alminderej FM. Study of new cellulosic dressing with enhanced antibacterial performance grafted with a biopolymer of chitosan and myrrh polysaccharide extract. Arab J Chem 2020; 13(2): 3672-81.
[http://dx.doi.org/10.1016/j.arabjc.2019.12.005]

[85]   Raju S, Manalel Joseph M, Kuttanpillai RP, Padinjarathil H, Gopalakrishnan Nair Usha P, Therakathinal Thankappan Nair S. Polysaccharide enabled biogenic fabrication of pH sensing fluorescent gold nanoclusters as a biocompatible tumor imaging probe. Microchim Acta 2020; 187(4)
[http://dx.doi.org/10.1007/s00604-020-4189-8]

[86]   Shoaib M, Quadri SMR, Wani OB, *et al.* Adsorption of enhanced oil recovery polymer, schizophyllan, over carbonate minerals. Carbohydr Polym 2020; 240: 116263.
[http://dx.doi.org/10.1016/j.carbpol.2020.116263] [PMID: 32475555]

[87]   Hutton B, Salanti G, Caldwell DM, *et al.* The PRISMA extension statement for reporting of systematic reviews incorporating network meta-analyses of health care interventions: checklist and explanations.

Ann Intern Med 2015; 162(11): 777-84.
[http://dx.doi.org/10 7326/M14-2385] [PMID: 26030634]

[88]   Moral-Muñoz JA, Herrera-Viedma E, Santisteban-Espejo A, Cobo MJ. Software tools for conducting bibliometric analysis in science: An up-to-date review. Prof Inf 2020; 29(1): 4.
[http://dx.doi.org/10 3145/epi.2020.ene.03]

[89]   Mao N, Wang MH, Ho YS. A bibliometric study of the trend in articles related to risk assessment published in science citation index. Hum Ecol Risk Assess 2010; 16(4): 801-24.
[http://dx.doi.org/10 1080/10807039.2010.501248]

[90]   Ellegaard O, Wallin JA. The bibliometric analysis of scholarly production: How great is the impact? Scientometrics 2015; 105(3): 1809-31.
[http://dx.doi.org/10 1007/s11192-015-1645-z] [PMID: 26594073]

[91]   Warlick SE, Vaughan KTL. Factors influencing publication choice: why faculty choose open access. Biomed Digit Libr 2007; 4(1): 1.
[http://dx.doi.org/10.1186/1742-5581-4-1] [PMID: 17349038]

[92]   Parmar A, Ganesh R, Mishra AK. The top 100 cited articles on Obsessive Compulsive Disorder (OCD): A citation analysis. Asian J Psychiatr 2019; 42: 34-41.
[http://dx.doi.org/10.1016/j.ajp.2019.03.025] [PMID: 30951931]

[93]   Bottle R, Hossein S, Bottle A, Adesanya O. The productivity of British, American and Nigerian chemists compared. J Inf Sci 1994; 20(3): 211-5.
[http://dx.doi.org/10.1177/016555159402000307]

[94]   Zyoud SH, Waring WS, Al-Jabi SW, Sweileh WM. Global cocaine intoxication research trends during 1975–2015: a bibliometric analysis of Web of Science publications. Subst Abuse Treat Prev Policy 2017; 12(1): 6.
[http://dx.doi.org/10.1186/s13011-017-0090-9] [PMID: 28153037]

[95]   Sweileh WM. A bibliometric analysis of global research output on health and human rights (1900–2017). Glob Health Res Policy 2018; 3(1): 30.
[http://dx.doi.org/10.1186/s41256-018-0085-8] [PMID: 30377667]

[96]   Sweileh WM. Global research output on HIV/AIDS–related medication adherence from 1980 to 2017. BMC Health Serv Res 2018; 18(1): 765.
[http://dx.doi.org/10.1186/s12913-018-3568-x] [PMID: 30305093]

[97]   Yataganbaba A, Ozkahraman B, Kurtbas I. Worldwide trends on encapsulation of phase change materials: A bibliometric analysis (1990–2015). Appl Energy 2017; 185: 720-31.
[http://dx.doi.org/10.1016/j.apenergy.2016.10.107]

[98]   Bakkalbasi N, Bauer K, Glover J, Wang L. Three options for citation tracking: Google Scholar, Scopus and Web of Science. Biomed Digit Libr 2006; 3(1): 7.
[http://dx.doi.org/10.1186/1742-5581-3-7] [PMID: 16805916]

SUBJECT INDEX

www.ingramcontent.com/pod-product-compliance
Lightning Source LLC
Chambersburg PA
CBHW050840220326
41598CB00006B/419